Are You Liv
Computer Simulation:
Elon Musk & Nick
Bostrom
Think It's Likely

Life After Death Book Three

By

Stephen Hawley Martin

WWW.OAKLEAPRESS.COM

www.OakleaPress.com

Below are Links to the first two books in this series:

Life After Death
Powerful Evidence You Will Never Die
Link: https://www.amazon.com/dp/B06W572NRB

Life After Death Book Two
Achieve Joy Now & Bliss in the Afterlife
Link: https://www.amazon.com/dp/B09GHY3SRN

CONTENTS

Chapter One: Bostrom's Theory

Are you living in a computer simulation? According to Nick Bostrom, a Swedish-born professor at the University of Oxford in England—a recipient of the Professorial Distinction Award with a background in theoretical physics, computational neuroscience, logic, and artificial intelligence—you and every other human being on Earth likely are. But like Neo in the movie, *The Matrix,* you just don't know it.

When Bostrom published a paper in 2003 that argued it's a strong possibility the reality we share is a computer simulation, he sent shock waves throughout the academic world. The idea has become a staple of popular culture. Many now think it's more than possible that we and our reality are the results of a computer simulation created by an advanced civilization. In 2016, for example, a *New Yorker* article reported that many in Silicon Valley had become obsessed with the idea, and that two unnamed billionaires had employed teams of scientists to attempt to break out of the simulation technology. Business magnate Elon Musk has gone public and acknowledged his personal fascination with this idea, including his conviction that the possibility that we are living in "base reality," i.e., one that is *not* simulated, is one in billions.

Many prominent individuals, scientists, and philosophers now apparently take seriously the seemingly bizarre possibility that we, and our world, are creations of a giant computer because, they say, it is only a matter of time before we will possess the technology to create such a digitally rendered reality. Over the past 50 years or so, we humans have created ever more realistic, virtual worlds. Proponents of simulation theory say that at present nothing stands in the way of these simulated realities continuing to improve *ad infinitum*. Some say they foresee a time when these virtual worlds will become so realistic that unless we have direct knowledge, in effect it will be impossible for us to tell the difference between a virtual world and base reality.

This raises the unsettling question, "How do we know that we don't already live in a simulation?" Could this have already happened? Could we be living in what Bostrom calls, "an ancestor simulation," a creation of an advanced human civilization designed for the purposes of studying its own past? He maintains that advanced civilizations may have countless reasons to run such simulations and would likely create astronomical numbers of them. It would therefore seem to follow that the overwhelming majority of conscious beings who may exist in this universe would exist in and inhabit simulations.

Bostrom's thesis has merit, and I believe that if it were not for a faulty premise, it could be true. In fact, it seems to me that in some ways it is very close to what I believe the truth to be about life in physical reality, and that is the likely reason so many intelligent and thoughtful individuals find it so compelling. Intuitively, much of what Bostrom lays out rings true to them. Nevertheless, as I will explain as you read ahead, his theory is off the mark.

If you have read the first two books in the *Life After Death* series and have considered the evidence presented in them dispassionately and thoughtfully, you will likely agree with me and come to the same conclusion. In this volume, I will set forth information not found in those books, and I will make an effort to keep references to evidence contained in the first two brief in an effort to avoid boring readers of those volumes.

Before I begin, however, let's review what Bostrom has proposed. He lays out three propositions and maintains that one of them must be true.

Proposition One, is that all technological civilizations collapse or go extinct prior to reaching technological maturity, and therefore we do not live in a virtual reality because, as a result, they do not exist.

Proposition Two is that civilizations can and do reach technological maturity, but for some reason, they never create simulations that resemble our world.

Proposition Three is that if we reject the first two propositions, we must accept his third and final proposition that we are almost certainly living in a simulation.

Let's take a close look at each of Bostrom's propositions.

Proposition One

The first proposition is that civilizations can never acquire technological maturity, and so they never get a chance to create what Bostrom calls, "ancestor simulations." It does seem logical that if civilizations other than ours exist, at least some of them would self-destruct. The story of Atlantis comes to mind. In our own case, we have developed atomic weapons, which give us the means to destroy civilization, and it seems quite possible it could actually happen. Due to their theological beliefs and hatred of Jews, the Iranian mullahs seem bent on creating Armageddon. Let's hope they never get the bomb.

Logic seems to indicate, however, that if civilizations are fairly commonplace throughout the universe, it would be unlikely that every one of them would self-destruct. After all, we live in a universe of hundreds of billions of galaxies that contain literally trillions of star systems, any one of which could be home to a burgeoning technological civilization. What could possibly prevent every single one

from advancing to the level of technological maturity required to create a convincing simulation?

I think the recent spate of UFO sightings argues
against the following conjecture, but some scientists believe no intelligent life exists in the universe, other than
our own. Proponents of this view draw from the observa-
tion known as the Fermi Paradox, which is that the universe around us seems oddly silent, and therefore, it may
be devoid of life other than on Earth. Yet, according to
many astrophysicists, the universe has been habitable for
life for a long time, beginning eons before our solar system
came into being. They tell us it's been around for 13.8 billion years and that Earth is only 4.6 billion years old. If
true, it would seem that more than enough time has
elapsed for advanced civilizations to develop somewhere
besides our planet. And yet, we see no signs of civilizations
other than our own.

How is that possible? Some scientists argue that the
evolutionary stages between inert matter and advanced intelligent life are so astronomically rare that they prevent
life from developing at all, much less life that evolves to a
high state of awareness.

Consider this: In 1953 James Watson [born 1928] and
Francis Crick [1916-2004] discovered the double helix, the
twisted-ladder structure of deoxyribonucleic acid (DNA),
which is necessary for life to exist. This marked a milestone

in the history of science, and it gave rise to modern molecular biology, which is largely concerned with understanding how genes control the chemical processes within cells.

Four years later, in 1957, Francis Crick realized that the chemical subunits along the interior of the double helix were functioning just like alphabetic characters in a written language, or the digital characters such as the zeros and ones in computer code, and that they are what direct the construction of proteins and protein machines that all cells need to stay alive. In other words, digital information directs the construction of the crucial components of living cells. Therefore, to explain the origin of life, one would have to explain how this complicated processing system came about.

Many, if not most scientists including Nick Bostrom, continue to hold fast to the basic premise of Scientific Materialism, which is that only material substance, i.e., "matter," exist. In other words, if it cannot be seen under a microscope, it cannot be. If that is true, intelligence and consciousness could not have come about until evolution produced a brain. Many scientists cling steadfast to this belief, even though after years of concentrated effort they have not been able to determine how the brain, i.e., inert matter, is able to do so.

This has created a conundrum that a professor at NYU, the well-known Australian philosopher and cognitive scientist, David John Chalmers, has labeled, "The Hard Problem." In a 2014 Ted Talk, Chalmers proposed a theory called panpsychism, which suggests that consciousness may be a fundamental building block of the universe like space, time, mass, and charge. Even a subatomic particle would have a tiny degree of consciousness. He went on to say that the level of consciousness an entity possesses might be linked to the amount of information being processed, which is why a human would have a higher level of consciousness than a dog or a cat or a piece of steel. Chalmers' theory actually supports Bostrom's theory in that a computer processing huge amounts of information would be conscious. Moreover, Bostrom maintains that a simulation's software will create the consciousness of those living in it, which is a point to keep in mind as we move ahead in this discussion.

The upshot is: If Scientific Materialist theory is correct, even with the addition of panpsychism, the DNA molecule would have to have come about by chance. If so, it would have done so in defiance of very long odds, the reason being that DNA can be compared to a computer program that is much more advanced that any humans have yet devised.

Just how advanced and complicated is it? According to an article on the website of *BBC Science Focus Magazine,* the UK's leading science and technology monthly: "The DNA in your cells is packaged into 46 chromosomes in the nucleus. As well as being a naturally helical molecule, DNA is supercoiled using enzymes so that it takes up less space. If you stretched the DNA in one cell all the way out, it would be about two meters long, and all the DNA in all your cells put together would be about twice the diameter of the Solar System."

How incredible is that? Think of the enormous amount of information packed into DNA—every cell in your body contains six and a half feet of computer code so small it takes a microscope to see it!

In case you want to verify that for yourself, here's a link to the article just referenced:

https://www.sciencefocus.com/the-human-body/how-long-is-your-dna/

Mathematicians who have calculated the likelihood of DNA happening by chance say the odds approach one in infinity. But even so, in a universe that literally contains trillions of solar systems, Bostrom argues that it is certainly possible that life has developed on other planets in addition to Earth.

If, however, life has only happened once in 14.8 billion years due to the incredibly long odds, one could argue that the overwhelming complexity of DNA actually suggests that we live in a simulation. It seems highly likely that some sort of intelligence must be behind the creation of life, as we know it. If, as Scientific Materials believe, that creator is not God or some other non-material, intelligent source, it could actually be an advanced civilization that came about by accident somewhere else, and we are living in a simulation that civilization created. We might even live in a simulation developed by our own descendents who have become technologically mature and are running an ancestor simulation. Alternatively, we could be living in a simulation created by artificial intelligence (AI) computers run amuck that our descendents created, as in *The Matrix* films.

Proposition Two

The second proposition is that civilizations can reach technological maturity and yet, in spite of this, they never author simulations that resemble our world. Such simulations could serve as advanced laboratories probing questions that would be inconceivable to minds at our current stage of development. In many respects, simulations would be the ultimate theory testers because the architects would have total control over all of the conditions within. Given the vast utility of such simulations, what plausible reasons could prevent our descendents from creating them? One possibility is that simulations that realistically resemble our world are impossible. Given the advancing realism of our own simulated worlds, however, this seems unlikely. So why take it seriously?

There are at least two reasons why true ancestor simulations might not exist. The first is that simulating a universe like ours would be computationally impossible because simulating every fine grain detail from subatomic particles to massive galaxies would be so computationally expensive that it would be unobtainable for even the most advanced technological society. The noted physicist Michio Kaku dismisses the simulation hypothesis for this reason. He maintains that the only information processor that is up to the task of simulating a universe is the universe itself. This may or may not be true, but it overlooks that a convincing simulation would not need to render every detail of the universe. Distant galaxies that are being observed, for example, could be compressed to just a few bits of information, much like in video games that render objects and scenes only when a virtual avatar is interactive with them. In this way, a simulation could dramatically cut computational costs by creating only those parts of the simulation that are of direct relevance at any given moment in time. Likewise, memories could be implanted in the entities inhabiting it to give the impression of continuity. In this way, if the architects wished to create a simulation that appears identical to our world, they would need only to manipulate the conscious experiences of its inhabitants.

This brings us to the second possible reason why such simulations might not exist. In The Matrix films, the human inhabitants of the Matrix remain connected to physical bodies that persist in base reality. However, as previously mentioned, Bostrom postulates that conscious inhabitants originate from within the simulation itself. There are reasons to believe that simulating a conscious mind digitally is impossible. We will return to this conjecture, but if we proceed under Bostrom's assumption that consciousness supporting simulations are possible, we must consider why an advanced civilization might choose not to create them. Broadly speaking, their objections may lie on ethical grounds that relate to the treatment of conscious beings.

Life is difficult for almost everyone in the reality that you and I inhabit today. Some individuals undergo terrible agony and suffering. Would highly intelligent beings in an advanced civilization be willing to cause such suffering to conscious, though less-developed, beings? Would they allow them to suffer and eventually cease to exist with no reward or benefit whatsoever to the being that endured that suffering? They might. After all, until they were declared an endangered species in 2015, we infected our closest cousins in the animal kingdom, chimpanzees, with diseases in order to study the effects of pharmaceuticals and medical procedures. We continue to do the same for other conscious beings such as baboons, cats, cows, dogs, ferrets, guinea pigs, hamsters, horses, llamas, and mice. We rationalize that the suffering we impose is justified because what is learned can be beneficial to us. Based on similar reasoning, we humans raise and slaughter pigs, sheep, cattle and other animals in order to eat them. Nevertheless, perhaps at the stage when advanced simulations become possible, societies will have put in place laws preventing this sort of thing.

If, however, there are many advanced civilizations throughout the universe as Bostrom suggests, would every one of them have such laws? Probably not. It would only take one advanced civilization to be in the business of creating simulations for there to be a huge number in existence at any given time.

Proposition Three

Bostrom's third proposition is that we are living in a simulation. He apparently believes that one of his three propositions is true and that the chances are about equal for each. Elon Musk thinks the chances that we are not living in a simulation is one in billions, and so he must be convinced that we are living in one.

I believe that in a way it is true we are living in a simulation, but it is not the sort of simulation that Bostrom and Musk have in mind. At least one thing about it is very different.In the next chapter, I will begin to explain what makes sense to me.

Chapter Two: A False Premise

Bostrom's propositions would be sound, and in my opinion we would have to accept his logic as well as his conclusion that we are likely living in a computer simulation if his theories were based on a valid premise. What I refer to is the basic assumption of Scientific Materialism that only material substance, i.e., "matter," exists. If true, consciousness would have to have been created at some point by a brain—there could be no other possibility. That being the case, Bostrom's belief that it is possible to create consciousness digitally would also certainly be possible.

The premise, however, is false. As will be demonstrated, matter does not create consciousness. Reality works the other way around. Conscious creates reality, and in turn, it creates matter.

In Book One of this series, *Life After Death, Powerful Evidence You Will Never Die,* a great deal of evidence was presented that indicates with no room for doubt I can see that the brain does not create consciousness, but rather, that the brain is a receiver of consciousness that integrates consciousness with the body. This is the conclusion researchers at the Division of Perceptual Studies (DOPS) have come to at the University of Virginia School of Medicine after almost sixty years of careful research and analysis. The evidence they cite falls into four categories:

1. Recovery of lost consciousness in the moments or days prior to death among people who have been unconscious for prolonged periods of time.

2. Complex consciousness ability in some people who have minimal brain tissue.

3. Complex consciousness in near-death experiences when the brain is not functioning or is functioning at a greatly diminished level. Those who have flat-lined on an operating table, for example, can often relate what doctors and nurses were saying or doing at that time, and they are able to describe the activities that were going on in the operating room, and sometimes in other locations as well.

4. Memories, particularly among young children, accurately recalling details of a past life. More than 2500 such cases have been studied since the early 1960s and well over half of them have checked out in terms of what the child remembered about the individual he or she claimed to have been, including names, occupations, locations, family members, manner of death, and so forth.

Since the brain does not create consciousness as the evidence to be found in Book One reveals, it will be impossible to create consciousness digitally with a computer. If you have any doubts about this and have not read the first book in this series, I suggest you do so as I do not wish to bore those who have read that book by regurgitating dozens of pages of details.

Matter Is Not What Most People Think It Is

In reality, matter is not what most people think it is—certainly not what the founders of Scientific Materialism thought it was back in the nineteenth century when the theory came to be. Matter is energy and energy is matter as in $E = MC^2$, according to which energy equals mass times the speed of light, squared. In other words, mass (matter) and energy are the same things in different forms—matter being tiny, rapidly vibrating subatomic particles. Moreover, all reality, the entire universe, is one single whole. Anyone wishing to expunge Scientific Materialist dogma from his or her thought processes ought to read Gary Zukav's book, *The Dancing Wu Li Masters.* He explains quantum mechanics without using complicated mathematics. Zukav writes:

> . . . *the philosophical implication of quantum mechanics is that all of the things in our universe (including us) that appear to exist independently are actually parts of one all-*

encompassing organic pattern, and that no parts of that pattern are ever really separate from it or from each other.

In the same book, Zukav also wrote:

The astounding discovery awaiting newcomers to physics is that the evidence indicates that subatomic "particles" constantly appear to be making decisions! More than that, the decisions they seem to make are based on decisions made elsewhere. Subatomic particles seem to know instantaneously what decisions are made elsewhere, and elsewhere can be as far away as another galaxy! The key word is instantaneously. How can a subatomic particle over here know what decision another particle over there has made at the same time the particle over there makes it? All the evidence belies the fact that quantum particles are actually particles.

As we will see, the reason those particles know is that the ground of being of reality is consciousness or mind. Let's take a look at an experiment that indicates this and also shows that consciousness or thought is able to create reality. It has to do with how light behaves, and it's called the "Double Slit Experiment." The strange but revealing phenomenon associated with it is known as "The Participating Observer." The findings of this experiment are straightforward, and have been replicated in many different laboratories. No honest scientist will refute them.

Scientists have known for more than a hundred years that light can behave both as waves and as particles (photons), but until 1905 they thought light was comprised only of waves. Thomas Young (1773-1829) demonstrated in 1803 that light is waves by placing a screen with two parallel slits between a source of light—sunlight coming through a hole in a screen—and a wall. Each slit could be covered with a piece of cloth. These slits were razor thin, not as wide as the wavelength of the light. When waves of any kind pass through an opening not as wide as they are, the waves diffract. This was the case with one slit open. A fuzzy circle of light appeared on the wall.

When both slits were uncovered, alternating bands of light and darkness appeared, the center band being the brightest. Scientists call this a zebra pattern. The areas of light and dark result from what is known in wave mechanics as interference. Waves overlap and reinforce each other in some places and in others they cancel each other out. The bands of light on the wall indicated where one wave crest overlapped another crest. The dark areas showed where a crest and a trough met and canceled each other out.

In 1905, Albert Einstein (1879-1955) published a paper that revealed light also behaves as though it consists of particles. He did so by using the photoelectric effect.

When light hits the surface of a metal, it jars electrons loose from the atoms in the metal and sends them flying off as though they were struck by tiny billiard balls. So, Thomas Young demonstrated light is waves, and Einstein demonstrated it is particles. This is the sort of paradox that led scientists to develop quantum mechanics.

Now let's consider a double slit experiment constructed to determine what happens when those conducting the experiment observe or do not observe which slits the photons of light pass through. This time a gun is used that fires one photon at a time. I first read about this experiment years ago in an article entitled, "Faster Than What?" in the June 19, 1995 issue of *Newsweek*. It reported on a paper to be published by a well-known quantum physicist, Raymond Chiao, then teaching at the University of California at Berkeley. Just so you'll know I'm not making this up, much later, in July 2008, I reviewed the facts of this experiment as they are presented here with a guest on the radio show I hosted at the time, the noted quantum physicist Henry P. Stapp, author of, *MINDFUL UNIVERSE: Quantum Mechanics and the Participating Observer* (Springer, 2007). He indicated I understood the facts correctly.

Both slits were open and a detector was used to determine which slit a photon passed through. A record was made of where each one hit. Only one photon at a time was shot, so one would suppose there could be no interference. This was the case. The photons did not make a zebra pattern. Rather, they made marks, tiny dots, on a screen.

When the detector was turned off, however, and it was not known which slit a photon passed through, the zebra pattern appeared. In other words, without the detector making it possible for the researcher to observe and know which slit particles passed through, the particles behaved like waves even though they were fired one at a time.

Imagine the stir this caused among those conducting the experiment. In the *Newsweek* article reporting on this, Nobel Prize winning physicist Richard Feynman (1918-1988) was quoted as saying this is the "central mystery" quantum mechanics, that something as intangible as knowledge—in this case, which slit a photon went through—changes something as concrete as a pattern on a screen.

But how could knowledge change the behavior of particles shot from a gun? Materialist science cannot produce an explanation because, as we know, a tenet of Scientific Materialism is that the brain produces consciousness, awareness, and thought, and that means consciousness, awareness, and thought must be confined within a person's skull. Since it would be ludicrous to suggest that thought enclosed inside a person's head could be capable of having an effect on photons shot from a gun, it ought to be clear to everyone who gives it a moment's thought that the tenet is false. Obviously, consciousness is not confined inside the skull. Yet Materialists tenaciously cling to the tenet, saying there must be two different sets of laws of physics: a small (subatomic world) set, and a macro world (the one we live in) set. Somewhere between these two worlds, the laws of physics must change.

The belief in two sets of physical laws does not explain why thought contained in someone's head should change things at the subatomic level, or anywhere else. Moreover, other experiments refute the contention that two different laws of physics exist. One such experiment involves large (carbon 60) molecules called "buckyballs," so big they can be seen and therefore are part of the macro world. Research shows they exhibit the same quantum properties as infinitely small particles. Another is an experiment conducted in 2002 that involved crystals. It produced quantum ridges half an inch high—large enough to be measured with a conventional macro-world ruler.

What makes more sense is what William of Ockham (c. 1287–1347) is thought to have been the first to say, "The simplest explanation is the best." The researcher's ability to know—his or her consciousness and mind—is what causes the waves to collapse into particles that form a pattern. In other words, when the researcher can access the knowledge, the zebra pattern does not occur. If he or she cannot access it, the zebra pattern appears. This was verified by setting up the experiment several ways. In the first, the detectors were in front of the two slits. In the second, researchers placed detectors between the screen and the

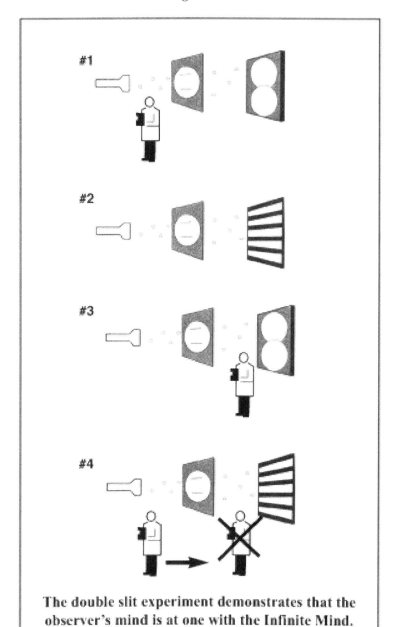

The double slit experiment demonstrates that the observer's mind is at one with the Infinite Mind.

two slits, i.e., after the photons had passed through them. As in the original experiment, knowing about a photon's behavior at the two slits made the zebra pattern vanish, whether or not the detectors were before or after the slits (see the accompanying graphic). But when the detectors were switched off, the zebra stripes returned.

In a third variation, a detector was placed before the slits and a mechanism erased the knowledge after the photon had passed through. The same thing happened. The zebra pattern returned. The result was the same no matter which way the experiment was set up—before the slits, after the slits, or before the slits and then erased. Whether or not the researcher was able to know where each photon hit determined the presence of the zebra pattern, or the lack of it. Versions of the experiment reported on in *Newsweek* were carried out at the University of Munich and at the University of Maryland. The behavior of the photons, the researchers report, "is changed by how we are going to look at them."

The question is, how? As mentioned above, the answer escapes scientists who refuse to think outside the Materialist box, but as soon as one is willing to step outside those four walls, the answer is clear. Consciousness—thought—creates reality. In the double slit experiment, it is the participating observer's consciousness that creates the resulting reality. Before the multitude of photons fired through slits is observed [seen or looked at], it exists as non-material potential in the form of waves. Observed, the waves collapse into something more solid—photons that form a pattern. What the researcher conducting the experiment thinks ought to be the result is the result—marks on a screen where the photons hit that he or she fired from the photon gun, and so the observer's mind creates the reality. If you can think of a more logical reason than that, please let me know. You can get in touch through my website: www.shmartin.com

How Physical Reality Is Created

I believe that mind creates reality, and I am about to provide evidence to support this contention, evidence that was collected by a college professor I spoke with on my radio show named Stephen Braude. At the time, Dr. Braude was a tenured professor of philosophy at the University of Maryland Baltimore County, and I had just read his book, *The Gold Leaf Lady and Other Parapsychological Investigations* (The University of Chicago Press, 2007).

The interview did not disappoint. Dr. Braude related several well documented and amazing stories of mind over matter, but perhaps the most fantastic, as well as the one that supports the contention that mind [Infinite Mind] creates matter, had to do with Katie, a woman born in Tennessee, the tenth of twelve children.

Katie is apparently a simple woman. Illiterate, at the time Dr. Braude wrote the book about her, she lived in

Florida with her husband and worked as a domestic. She was also a psychic who'd had documented successes helping the police solve crimes. In one instance she was able to describe the details of a case so thoroughly and accurately, the police regarded her as a suspect until those actually responsible were apprehended. She apparently was also able to apport objects—in other words, she somehow caused them to disappear in one place and reappear in another, at least that is what Dr. Braude maintained when I spoke with him. And that wasn't not all. Seeds reportedly germinated rapidly in her cupped hands. Observers claim to have seen her bend metal, and she was both a healer and a medium or channel. Being illiterate, she cannot read or write in her native English, but she has been video taped writing quatrains in medieval French similar both in style and content to the quatrains of Nostradamus.

I know some scientists are going to flip out when they read what comes next because it goes against what they see as a fixed law of physics—that matter cannot be created nor destroyed—but most amazing, perhaps, is what appeared spontaneously on her skin—on her hands, face, arms, legs, and back—apparently out of thin air. It looked like gold leaf, a thin version of the wrapping on a Hersey's Kiss. Katie could not control when this happened, but Dr. Braude and other witnesses saw the foil materialize firsthand. He even videotaped it appearing on her skin.

I just stopped typing and checked. As of this writing, footage from this video can be seen on YouTube. Go to YouTube and put "gold leaf lady Braude" in the YouTube search bar. Several videos about the Gold Leaf Lady will appear. In the search I just made, the one I am referring to was the third one down. The title of it is "UMBC In the Loop: Stephen Braude."

Dr. Braude took the foil to be analyzed. It turned out not to be gold at all, but brass—approximately 80 percent copper and 20 percent zinc.

Dr. Braude thinks there's a reason she produces brass and not gold. Where does the brass foil that appears on Katie's skin come from? It appears that her mind creates it. In fact, as mentioned, Dr. Braude believes she produces brass rather than gold for a reason. You see, at the time Katie has a difficult and tense relationship with her husband. Once she aported a carving set. It just appeared. And her husband—apparently nonplussed—said, "So what? It's not worth anything." Soon afterward, gold colored foil began appearing on Katie's skin. But it wasn't real gold, it was fool's gold—brass. Dr. Braude thinks this is how she gets back at her husband. Katie's mind—albeit the unconscious part—creates matter in the form of brass foil. This being the case, why should it be difficult to believe that an Infinite Mind—one infinitely more powerful than a human mind—created the material universe? The physical universe had to come from something.

As previously discussed, one thing quantum physicists agree on is that the physical universe isn't really physical. It is energy—vibrations—in other words, waves. In the double slit experiment, the researcher's ability to know turned waves into photons, which is to say that thought or mind turned waves into things [particles]. It does not seem to me too big a leap to suggest that mind is what creates the world we live in.

Chapter Three: A New Theory of Everything

Just over 500 years ago, people in the western world thought the earth was flat, that it was located at the center of the universe, that the sun and stars revolved around it, and an anthropomorphic God had created it. Then along came Christopher Columbus [1451-1506], Ferdinand Magellan [1480-1521], Johannes Kepler [1571-1631], and Galileo Galilei [1564-1642], and beliefs changed. The earth was now thought to be round. The sun was thought to be at the center of the universe, and an anthropomorphic God created it.

This remained what people thought until 1859 when Charles Darwin [1809-1882] published *On the Origin of Species.* After the contents of that book were widely known, if you were in academia or were highly educated, the only socially acceptable things to believe were that the universe had always existed and life had come about by accident. Material substance—matter—was all there was, and intelligence and consciousness did not, and could not have existed, until evolution produced a brain. This remained the scientifically accepted worldview well into the twentieth century. The universe was thought to be much smaller than we know it to be today, it was thought to have always existed, and the sun was believed to be at the center of it. Scientists believed that the Milky Way, what we now realize is one of countless galaxies, was all there was to the universe. Moreover, highly educated individuals still believed life had come about by accident and that intelligence and consciousness did not exist until evolution produced a brain.

Then Edwin Hubble [1889-1953], for whom the Hubble Space Telescope is named, changed how reality is viewed once again by discovering and reporting that countless galaxies exist in addition to Milky Way, that the universe is expanding, and that the sun is definitely not at the center of it. Hubble published his first paper on the relationship between red shift and distance in 1929, again revolutionizing our understanding of the universe and our place in it—although many continue to believe today that intelligence and consciousness did not exist until evolution produced a brain.

Beliefs are, however, in the process of changing once again. A major revolution in our understanding of the universe is occurring, and it is my hope that this book will inspire at least a few intrepid scientists to work out the mathematics, or whatever calculations are deemed necessary to support the new cosmology this book suggests, i.e., a new theory of creation of the universe and life sufficient to convince diehard Scientific Materialists that intelligence and consciousness were in existence long before evolution produced a brain, and that they in fact are the Source of the universe and life.

Much of the work has already been done. John Samuel Hagelin, a Harvard-educated Ph.D. in Quantum Physics, has pointed toward such a theory by comparing the unified field, which many scientists speculate is what existed before the Big Bang, to the field of pure consciousness known as "Veda" described in the ancient religious texts of India called the *Vedas*.

Below is a link to a YouTube video of a lecture by Dr. Hagelin about this:

https://www.youtube.com/watch?v=4u3f7_p1i8c&t=972s

Dr. Hagelin maintains that the unified field and Veda are one and the same. Here are two paragraphs of text quoted verbatim from the write up under the aforementioned video:

> *Two sciences, one ancient and subjective and the other modern and objective, describe manifest creation as an expression of infinite dynamism embedded in the infinite silence of the underlying field. Physics describes this relationship in terms of the unified field and vacuum energy, and Vedic Science in terms of Shiva and Shakti.*

Similarly, at every level of manifest creation, these two descriptions of nature's functioning correspond exactly. Physics describes three super fields giving rise to the five spin types that characterize elementary particles, the resonant frequencies of the unified field and building blocks of creation; Vedic Science speaks of the three Prakritis or Doshas giving rise to the five Mahabhutas that structure the universe.

Allow me to cut to the chase and do my best to put into simple, everyday language the new theory that I believe coincides with and is supported by the statements above. I say this because the ancient Rishis of India—sages who lived thousands of years ago—believed that Veda, which boiled down into everyday language is *consciousness,* is the ground of being of all that is. In other words, the Rishis thought consciousness gives rise to the universe.

This book was written to impart and explore a theory that agrees with Dr. Hagelin that the Rishis were correct. Moreover, the theory supports the Rishis' belief that Veda (consciousness) is the core of each human being and every living thing.

In humans, Veda, or consciousness, manifests as the Self—the "I AM" or the silent observer at the back of your mind and my mind. But do not misunderstand. The Self (Veda or consciousness) is not an individual's mind. Veda, the ground of being, is what directs your hand to turn a page of this book. It is the decision-maker in each of us that observes the thoughts that arise in the mind and decides which ones to act upon and which ones to dismiss.

It is impossible to understate the magnitude of this realization, the implications of which are huge. A few will be touched upon in upcoming chapters.

What Existed Before the Big Bang?

Does any of this fit with what the majority of scientists believe today?

Physicists have proposed that the spark of existence had its origin in a quantum fluctuation, triggering an explosive chain reaction—the Big Bang—leading to the still evolving universe we inhabit today. This narrative, however, presupposes the laws of quantum mechanics. As British Biochemist Rupert Sheldrake said in a now banned TED Talk, "[Scientists today say] give us one free miracle and we'll explain the rest.' And the one free miracle is the appearance of all the matter and energy of the universe, and all the laws that govern it, from nothing in a single instant."

Suffice it to say that rather than explaining existence, current scientific theories of the origins of the universe have simply pushed things back to a point that raises the question, "What existed before the beginning?" Could it all have come from nothing? Although that is what many scientists purport to believe, it doesn't make sense. As the song in *The Sound of Music* goes, "Nothing comes from nothing, nothing ever could."

Instead of beginning with nothing, it seems logical that the challenge of explaining existence should focus instead on defining a self-existing ground of being for which no explanation is required. Some physicists have proposed that the true ground floor of reality is the seething quantum realm of particles, forming in and out of existence. While this level of reality no doubt exists, there is no clear reason why the primordial situation should be constrained by quantum physics. A deeper level of explanation seems to be required, and what makes sense to me, as mentioned above, is that consciousness is the ground of being. How seething quantum particles came to be the ground of reality calls out for an explanation, but consciousness can explain itself. A unique feature of consciousness is that it does not appear grounded in anything beyond itself. The conscious self is self-producing in so far that it exists only in and to itself. As René Descartes [1596-1650] famously said, "I think therefore I am." In other words, nothing is required beyond consciousness for existence to be a demonstrated fact.

Diehard Scientific Materialists will have a hard time giving up the idea that the brain is what creates consciousness, but anyone familiar with the research by the University of Virginia touched upon at the beginning of Chapter Two knows that's not the case. It's a shame the findings of that research have not been widely publicized—I suspect because science journalists are afraid of being ridiculed by Materialists—but as previously discussed, the brain is a receiver that integrates consciousness with the body, which might better be called the "mind-body complex" because it is a vehicle or apparatus that allows your consciousness to inhabit and operate in this (physical) dimension. If you don't believe this, go to YouTube and put the following phrase in the search bar: "Dr Bruce Greyson Consciousness Independent of the Brain." A lecture by Dr. Greyson that goes into detail about UVa's research should appear at or near the top. Dr. Greyson, by the way, is not some New Age Looney Tune. He is Professor Emeritus of Psychiatry and Neurobehavioral Sciences at the University of Virginia School of Medicine.

All Is One

If the brain does not create consciousness, where do you suppose it comes from? As stated above, consciousness is the Source, the Veda or ground of being that generates and informs reality. Let's consider evidence that supports this. The British quantum physicist, Sir James Jeans [1877-1946], wrote the following: "The universe begins to look more like a great thought than a great machine." He was not only onto something, he hit the nail on the head. He knew what all quantum physicists know, that there is no such thing as matter per se—no separate things made of truly solid stuff. All that exists are vibrations—energy, pure and simple. And since physical reality is vibrations, nothing is truly separate—everything is connected. There can be no barrier or edge where one vibration stops and another one begins. As mystics have been saying since the dawn of time, "All is one."

Moreover, if consciousness is primal, consciousness must create everything, including the building blocks of space, time, mass, and charge. As Sir James Jeans seems to have suggested above with his comment about the universe resembling a thought, the physical reality we inhabit might be compared to a thought or a dream the Source is having.

Consider this. When you and I dream at night, our minds create our dreams, their imagery and all their trappings, and yet they seem completely real. If a car in your dream is about to run you over, for example, you are certain it's going to hurt, which is why you wake up. Yet not until your eyes pop open, do you realize it was just a dream.

Now ponder this. There has to be a character in your dream from whose point of view what's happening is observed. There cannot be a dream without someone observing it, and in your dreams that character is you. It follows that if the universe and this world is a "dream" in the mind of the Creator, there has to be a character in the dream for it to be observed. Who is the character?

You are that character, and I am that character.

We humans—all living things—are characters in the Creator's dream that observe what's going on. And since all is one and connected, including the Creator and us, we serve as the Creator's eyes and ears. We are the vehicles by which the Source knows his/her/its creation.

The reason this may be difficult to accept is that in our culture we are used to thinking of ourselves as totally separate entities. We think of the universe as "out there." You think of this book, or the Kindle device or phone you are holding, as a separate object. But as noted above, quantum mechanics says that isn't the way things really are. We are all one, all part of the dream. This will begin to sink in and make sense if you do not dismiss it out of hand and truly think about it, objectively.

Panpsychism Is Off the Mark

Let me say here parenthetically that the theory being put forth in this book differs completely from the one called "panpsychism," which some Scientific Materialists are promulgating in an attempt to explain how matter creates consciousness. Panpsychism is the idea touched on in Chapter One that was put forth in a 2014 Ted Talk by David John Chalmers. It is that everything material, however small, possesses an element of consciousness—that at the core, even an atom or a quark, photon, or an electron has a tiny level of consciousness.

According to this theory, consciousness is linked to information processing—the more information being processed, the higher the degree of consciousness. This would explain why a human has a higher level of consciousness than a mouse or an ear of corn. Based on this theory, complex information processing activity would in turn create complex consciousness. This actually supports Bostrom's theory, because a computer processing huge amounts of information would, therefore, be conscious.

However, based on research and observations by DOPS researchers at the University of Virginia School of Medicine, the panpsychism theory cannot be valid. To repeat briefly what the UVa researchers have found, the brain is a receiver of consciousness, it does not create it. How else could children remember past lives, given that between lives they had no brain? How could people be conscious of what doctors and nurses are talking about, and how could they observe what is going on in the operating room, when their brains have flat-lined and they are clinically dead? When the brain has flat-lined, it cannot be processing information.

Only Consciousness Is Conscious

It is also important to understand that in an attempt to find a solution to the mystery of creation, Materialists have been attacking it from the wrong direction. Physical reality is generated by and contained within consciousness, not the other way around. Matter does not create consciousness. Consciousness creates matter, and therefore, consciousness is the place to begin an investigation.

To get one's mind around this, it may be helpful to realize that it is impossible for anyone or anything to experience physical reality directly, which is an indication physical reality does not exist and cannot exist outside of consciousness. We experience this dimension through sight, hearing, touch, smell, and taste—the five physical senses of our body-mind complex. We cannot experience any physical thing without using at least one of those senses. Moreover, our consciousness and the Source—the Veda—are one and the same—the "I AM" in each of us.

Only one consciousness exists, which is the Source, and since the Source is all that is, it cannot step outside itself to observe itself. But the Source, aka Infinite Mind and intelligence, has found a way. We are the vehicles by which the Source experiences its creation.

Chapter Four: Implications of the New Theory

Understanding that consciousness is the ground of being, and that the true Self in each of us is that ground of being will be a boon to humanity once it becomes widely known and accepted. Moreover, it can be a boon to you today and to others who do not wait to take it to heart. The undeniable implication of the reality that all minds are connected and at one with the Source is that all humans are inherently equal, and that our thoughts and beliefs are what create our personal realities. Skin color, national origin, and cultural backgrounds do not have to hold anyone back. In a free society, the only victims will be those who consider themselves to be victims. Understanding that your consciousness is not isolated or trapped within your skull will prepare you and others to receive and accept knowledge that will open doors you may not have realized were there. Think about it. You are the Source experiencing your creation. Your mind and the mind of the Source are connected and influence one another. You have free will and can go with the flow created by the Source, or you can rail against it. The choice is up to you. Almost everyone alive today is blind to this because it doesn't fit into the worldview they hold. Hopefully, this book will help many who read it to open their eyes.

To illustrate how our mental framework can blind us, consider what Charles Darwin found on his visit to Micronesia during his voyage on the Beagle. At that time, the natives of those islands were so isolated from the rest of the world that they had never seen a ship. Darwin and others from the Beagle came ashore in dinghies. The natives had no difficulty seeing them. After all, they themselves used small boats. But they did not, apparently could not, see the Beagle moored offshore—even when it was pointed out to them. A boat of its size did not fit into their mental framework. As a result, it was invisible to them. The same is true today, for example, when it comes to ill health and aging. Then, as now, the way that most people see the world is what keeps it hidden. Some who read what is written here at first may not "see" what I'm talking about. All I ask is that you suspend disbelief as you read ahead.

In the spring of 2000 a startling realization came to me after a local radio interview about one of my novels. It was evening. I was beat, having just spent an intense hour trying my best to be entertaining and witty. On my way home, I stopped at my local Seven Eleven for a bottle of beer. A sign caught my eye as I approached the register.

"We I.D. under 27 years of age."

I took my place in line behind a couple of teenagers with Slurpies. An acquaintance from college took the spot behind me, and we exchanged pleasantries. My turn came, I put the bottle on the counter and reached for my wallet.

The clerk eyed me. "Sorry, I'll have to see your I.D.," she said.

"Excuse me?" I said.

"I'm going to need to see your I.D.," she repeated.

"You're kidding," I said.

She let out an exasperated sigh. "No, I need to see your I.D. before I can sell you that beer."

I placed my driver's license in her hand, turned to my friend, and gave a little shrug. Her mouth gaped. "It's true," she said, shaking her head. "You really do look young."

On the way home, I sipped, kept an eye out for police, and pondered the fact that I'd been asked to prove I was old enough to buy alcohol. You see, I was fifty-five years old at the time—more than twice what the clerk was required to I.D. It's definitely true that I felt much younger. Even today, almost twenty years later, I can detect almost no difference in how I feel now and how I felt when I actually was twenty-seven.

After that encounter, I started wondering why I appeared so young, and after a while, a possibility surfaced in my mind. Thirty years before, when I was 25, I'd read an article about a study of people who'd been consuming large doses of vitamin E for ten years. The article said that no measurable signs of aging had occurred among them. So I went out and bought a bottle, and I've been taking it since.

For years, I believed I wouldn't age. And for years, it seemed I didn't age.

Much later, I read that researchers had concluded that vitamin E in pill form cannot be proven to retard aging. As has often been the case, newer studies refute older ones. But I kept taking it anyway.

According to recent articles, we've come almost full circle. No researcher is ready to say vitamin E stops aging altogether, but new research indicates that taking the vitamin results in lower incidence of heart disease and cancer, while helping mitigate all sorts of health problems. Even so, I've come to believe that back then it may have worked for me in large measure due to the placebo effect. But it worked. Thirty years before I had read an article that said I wouldn't age if I took it. I expected it to work, so it did. If the following week I'd read another article that said the anti-aging qualities of vitamin E were hogwash, I probably would not have experienced the same result.

Belief is extremely potent. The effectiveness of placebos has been demonstrated time and again in double blind scientific tests. The placebo effect—the phenomenon of patients feeling better after taking dud pills—is seen throughout the field of medicine. One report says that after thousands of studies, hundreds of millions of prescriptions and tens of billions of dollars in sales, sugar pills are as effective at treating depression as antidepressants such as Prozac, Paxil and Zoloft. What's more, placebos cause profound changes in the same areas of the brain affected by these medicines, according to this research. For anyone who may still have been in doubt, this proves beyond a doubt that thoughts and beliefs can and do produce physical changes in our bodies.

In addition, the same research reports that placebos often outperform the medicines they're up against. For example, in a trial conducted in April, 2002, comparing the herbal remedy St. John's wort to Zoloft, St. John's wort fully cured 24 percent of the depressed people who received it. Zoloft cured 25 percent. But the placebo fully cured 32 percent.

Taking what one believes to be real medicine sets up the expectation of results, and what a person expects to happen usually does happen. This book will explain why. It's been confirmed, for example, that in cultures where belief exists in voodoo or magic, people will actually die after being cursed by a shaman. Such a curse has no power on an outsider who doesn't believe. The expectation causes the result. If you've read my novel, *The Secret of Life: An Adventure Out of Body, Into Mind,* you know I used this phenomenon as a factor in the plot.

Belief is powerful. It is the key to manifesting your desires. A study carried out on the Discovery TV Channel, for example, gives an indication. In this case, two researchers conducted the same ESP experiment in the same laboratory using the same equipment. They went to great pains to keep everything identical except for one thing. One researcher believed ESP was valid and the other did not. Both tests were supervised by impartial observers, including the Discovery Channel TV crew.

The experiment that employed the researcher who believed in ESP had a statistically significant number of correct scores, indicating the experiment was a success. The validity of ESP was demonstrated scientifically. But the correct hits in the experiment with the doubting Thomas researcher were within parameters that could be accounted for by chance, meaning the experiment failed to demonstrate the validity of ESP. Apparently, the one and only variable—belief—made the difference. The first researcher believed and the second didn't. Each got the result he expected.

The same thing is at work in prayer by believers. Prayer works. Prayer is thought released into the subconscious. Prayers give spirit, or the Life Force, extra zest that bolsters its natural tendency to organize matter in a way that is beneficial to life. Soon it will be clear to you precisely how this works, and in Chapter Nine we will cover in some depth the effects of prayer as demonstrated in scientifically-constructed, double-blind experiments.

We indeed create our own reality. How this happens is explained in lectures I came across years ago by a man named Thomas Troward. He first delivered them at Queens Gate at Edinburgh University in Scotland in 1904. Called *The Edinburgh Lectures on Mental Science,* they provide a clear-cut and plausible explanation that fits perfectly with the findings of studies on prayer—that distance between those praying and the one being prayed for is not a factor, and that the one being prayed for does not have to know about the prayers on his or her behalf. How prayer works is simple, but let me lay some groundwork before I place it before you.

It helps to begin by considering the difference that appears to exist between what we think of as "dead" matter and something we recognize as alive. A plant, such as a sunflower, has a quality that sets it apart from a piece of steel. The sunflower will turn toward the sun under its own power. When first picked, it possesses a kind of glow. This quality might be called the Life Force, or spirit. On the other hand, the piece of steel appears totally inert. Yet, at the quantum level, the steel is alive with motion. In fact, quantum physicists tell us that motion or energy is what comprises all matter. Atoms and molecules are not solid things. They are energy. Vibrations. Some would say the whole universe is alive, as though it were a single giant thought—the thought of an infinitely vast mind of organizing intelligence.

Even so, by outward appearances the sunflower is alive, and the steel is not. Few would argue this. But one might argue that a plant's state of "aliveness" is different from an animal's. Consider the difference in aliveness between a sunflower, an earthworm, and a goldfish. Each appears to be progressively more alive.

Now, let's add a dog, a three year old child, and a stand up comedian on a late-night talk show. Each has a progressively higher level of intelligence. So, to some extent, what we call the degree of "aliveness" can be measured by the amount of awareness or intelligence displayed—in other words, by the power of thought.

As has been written above, intelligence, or thought, underlies and creates the entire universe. But it becomes more evident to us—we can see it more clearly—as this intelligence becomes more self-aware. So the distinctive quality of spirit, or life, is thought, and the distinctive quality of matter, as in the piece of steel, is form.

Consider for a moment form versus thought. Form implies the occupation of space and also limitation within certain boundaries. Thought (or life) implies neither. When we think of thought or life as existing in any particular form we associate it with the idea of occupying space, so that an elephant may be said to consist of a vastly larger amount of living substance than a mouse. But if we think of life as the fact of "aliveness," or animating spirit, we do not associate

48

it with occupying space. The mouse is quite as much alive as the elephant, notwithstanding the difference in size. Here is an important point. If we can conceive of anything as not occupying space, or as having no form, it must be present in its totality anywhere and everywhere—that is to say, at every point of space simultaneously.

Life/thought not only does not occupy space, it transcends time. The scientific definition of time is the period occupied by a body in passing from one point in space to another. So when there is no space there can be no time. If life/thought is devoid of space, it must also be devoid of time. The bottom line is that all life, or thought, must exist everywhere at once in a universal here and an everlasting now as scientists who have studied this and worked it out mathematically would agree.

How does this help us understand how we create our own reality as well as how prayer works?

First, it is implicit in the discussion above that there are two kinds of thought. We might call them lower and higher, or subjective and objective because what differentiates the higher from the lower is the recognition of self. The plant, the worm, and perhaps the goldfish possess the lower kind only. They are unaware of self. Perhaps the dog, and certainly the boy and the comedian possess both. The higher variety of self-aware thought is possessed in progressively larger amounts as if ascending a scale.

49

The lower mode of thought, the subjective, is the sub-conscious intelligence or mind present everywhere that, among other things, supports and controls the mechanics of life in every species and in every individual. It causes the plant to grow toward the sun and to push its roots into the soil. It causes hearts to beat and lungs to take in air. It controls all of the so-called involuntary functions of the body. And, as we will see, it controls a lot more.

Levels of Mind

Subjective, Non-Dual Ground-of-Being Mind
The Collective Subconscious Mind
An Individual's Subconscious Mind or Soul
An Individual's Unconscious Mind
An Individual's Conscious Mind

There may actually be more levels of mind than shown above, but I have included this simplified chart to help the reader understand the accompanying text. I say there may be more because according to teachings of the College of Metaphysics in Missouri, there are actually seven different levels of mind, not just five.

That this lower kind of thought is everywhere at once coincides with the theory of Carl Jung who maintained that we humans share a universal mind. Moreover, we each have our own portion, our individual subconscious mind that blends into a shared collective mind that contains archetypes and so forth, which in turn blends into the ground-of-being, subjective mind referenced above. Our conscious minds are the producers of thought that make us self-aware. The various types of mind are inextricably linked, in that they arise out of the subjective, ground-of-being mind that is non-dual. Transcendent and non-dual, this ground-of-being mind doesn't distinguish between good and evil. It is what existed before humans evolved to a level that they could figuratively step out of themselves and consider what was "good" and what was not good. That point of time in evolution is recounted metaphorically in the story of Adam and Eve when the couple ate the fruit of the tree of the knowledge of good and evil and realized they were naked. As Shakespeare's Hamlet said, "Why, then, 'tis none to you, for there is nothing either good or bad, but thinking makes it so." The gradual emergence of self-aware thought out of this non-dual, subjective mind is implicit in our consideration of the plant, earthworm, goldfish, dog, boy, comedian and so forth up the scale.

Now let us consider an important point made in the lectures. The conscious mind has power over the subjective mind that creates our reality. I discovered the truth of this firsthand in college when I learned to hypnotize others. I would put a willing classmate into a trance and tell him he was a chicken or a dog. Much to the amusement of my audience, he would then act accordingly.

Hypnotism works because the hypnotist bypasses his subject's conscious mind and speaks directly to the subject's subjective mind. This part of the subject's mind has no choice but to bring into reality that which is communicated directly to it as fact by a conscious mind. Being totally subjective, it cannot step outside of itself and take an objective look. As such, it is capable only of deductive reasoning, which is the kind that progresses from a cause (the conscious mind's directive) forward to its ultimate end, having the mind of a golden retriever. It does not stop to question or analyze. This is the reasoning that a criminal might use in committing a crime. He may walk into a room, see a man counting his money, and think: "I need money, so I will take his. Since the man is protecting the money, I will get rid of him. I'll shoot him. He'll drop to the floor. I will then take the money and run. I'll leave by the window." Right and wrong, good and bad, aren't considered, only how to get to the end result.

On the other hand, the conscious mind, being objective and self-aware, can step outside. It can reason both deductively and inductively. To reason inductively is to move backward from result to cause. A police detective, for example, would arrive at the crime scene and begin reasoning backward in an attempt to tell how the crime was committed and who might have done it.

The result is that the subjective mind is entirely under the control of the conscious (objective) mind. With utmost fidelity, the subjective will work diligently to support or to bring into reality whatever the conscious mind believes to be true. Since the individual's subjective mind blends with the ground-of-being mind and is present everywhere, it is able to influence circumstances and events so that whatever the conscious mind believes to be true will indeed become true. So, for example, if I believe I am a sickly person, I will be a sickly person. If I believe that by sitting in a draft I will catch a cold, I will catch a cold when I sit in a draft. Conversely, if I believe that I am rich, that I deserve to be so because it is my birthright, I will become rich even if this is not already the case. If I think I am unlucky, I am unlucky.

This also explains how, why and when prayer works. When people who pray sincerely believe their prayers will have a positive effect, their prayers most certainly will. The belief they hold is impressed upon their own subjective minds. Their subjective minds blend into the ground-of-being mind. The more people praying and believing, the greater the effect. The subjective mind of the person for whom they are praying is also part of the ground-of-being mind, and the latter goes to work to bring about positive results.

It does not make any difference whether the person praying is at someone's bedside or halfway around the world. As noted above, thought, and therefore prayer, is present everywhere at once. It is nonlocal. This explains why prayer is not hindered by distance.

Most people go through life hypnotized into thinking that they have little or no control over their circumstances. The fact is that they create their circumstances with their thoughts and beliefs. The message of the Edinburgh Lectures is simple. Change your beliefs and your circumstances will change. And while you are at it, a few well-intentioned prayers can't hurt.

Returning to my personal experience with the power of belief and aging, as the days went by after that episode in the Seven Eleven, something else occurred to me. Starting a new life can have the result of deducting decades from a person's chronological age. Some years before, in 1993, I realized I'd gone stale in my career. I began having a recurring vision of myself coming around the track again and again. You might call it daily deja vu. I'd excelled in my career—was president of my own advertising agency. I was pulling down a salary well into six figures, was listed in *Who's Who in the Media and Communications.* I'd done what our society and our educational system seem to indicate should be the primary goal of life and the one true way to happiness and fulfillment. I'd picked a profession and risen to the top.

And as many who reach such rarefied air also have found, it wasn't all it had been cracked up to be.

Don't get me wrong. I love the creative process, and being creative is what advertising is all about. But when you are successful, and this is true in many lines of work, after a certain point you often end up no longer doing what made you successful. You end up supervising others who get to have the fun, and you get the headaches. So I sold my ad agency and took a year off before I began doing some marketing communications work again because, by that time, it was apparent I needed money feed my family, put a roof over our heads, and pay tuition bills. Nevertheless, I arranged my advertising and marketing communications work so that I was able to do the part I like, which is creating ads and campaigns, but also have time to do what I really and truly wanted. You see, back in 1993, that something was calling me.

I have since found that when you want something, really want it and remain attentive, an opportunity will appear. The late Joseph Campbell [1904-1987] labeled this opportunity "The call to adventure." This call will come whether the desire you hold is known to you on a conscious level, or whether it's hidden in your subconscious mind or soul. You'll be presented with a choice. You can follow your adventure and gain from it. Or you can refuse the call, in which case you will stagnate and eventually die—figuratively, or perhaps even literally. This points to a cause of much ill health that's hidden beneath our noses. Here is an important truth: To accept the call to adventure is to choose life over old age and death.

Myths of all cultures recount the same tale repeatedly, each in its own cultural guise. This is not surprising since the call to adventure is something each of us receive, often many times during a lifetime. We are compelled to leave the safety and security of our home base, what Campbell called the hero's "Ordinary World," and venture forward into the unknown where inevitable dragons and demons of one kind or other must be faced and overcome. Supported by unseen or supernatural powers, the hero who pushes forward will invariably succeed, later to return to familiar territory more highly evolved than when he or she left, and in possession of a new level of understanding. You see, the goal of life is evolution, which is why we leave our true home, the non-physical realm of spirit, and come to this rough and tumble place called Earth. It is to experience the good and the bad, to face hardships and overcome them, because that is how we grow and evolve. Our society by and large is unaware of this, even though the call comes whenever the time is at hand to move to a higher plane of understanding. This denial of such a basic component of life is particularly tragic in that dire consequences always result from our refusal to accept the call.

You need not take my word for this. Warnings can be found in myths throughout the ages. Refusal converts what otherwise would be positive and constructive into negative form. The would-be hero loses the power of action and becomes instead a victim bound by boredom, hard work, or even imprisonment. King Minos refused the call to sacrifice the bull, for example, which would have signified his submission to the divine. Of course, he didn't know that this would have resulted in his elevation to a higher state. So instead, like a modern day business executive or professional, he became trapped by conventional thinking and attempted to overcome the situation through hard work and determination. Indeed, he was able to build a palace for himself, just as many executives and professionals today build their mansions in the suburbs. But it turned out to be a wasteland, a house of death, a labyrinth in which to hide, and thus escape from the horrible Minotaur.

And look at what happened to Daphne, the beautiful maiden pursued by the handsome Greek god, Apollo. He wished only to be her lover, and he called to her, "I who pursue you am no enemy. You know not from whom you flee. It is only for this reason that you run." All Daphne had to do was submit, to accept the call, and beautiful and bountiful love would have been hers. She, too, would have had a relationship with the divine. But as you probably know, she did not submit. She kept running, and as a result turned into a laurel tree, and that was the end of her.

Two stories come to mind from popular culture that illustrate what I'm hoping to get across about answering the call and learning from the adventure. The first is *The Wizard of Oz*. In this one, Kansas is the heroine's (Dorothy's) "ordinary world," where we find her pining away about a better place that she believes might be, "somewhere over the rainbow."

Along comes Miss Gulch who wants to have her dog, Toto, destroyed because Toto has nipped at her. Miss

Gulch takes Toto away for that purpose, but the dog escapes and comes home. Dorothy runs away with Toto but eventually returns home, only to be swept away from the safety and security of the farm where she lives by a tornado. This takes her to an unfamiliar, unknown place, the Land of Oz, where she makes friends, faces and overcomes obstacles, and figuratively enters the belly of the whale in the form of the witch's castle. Then, when things look hopeless, Dorothy succeeds in what she has set out to do—she captures the witch's broom. Finally, even though the wizard turns out to be a fraud, which creates what storytellers call "the dark moment," Dorothy is able to return home.

As the story ends, we find that she has evolved as a result of her experiences and adventure, and that she now processes an important understanding she did not before. Dorothy now knows that there is "no place like home."

The second story is in my opinion an allegory of what life of Earth is about. I'm referring to the movie, *Groundhog Day.*

At the beginning of the story, the hero, Bill Murray's character, and a TV crew have left Pittsburgh, their ordinary world, and have journeyed to Punxsutawney, Pennsylvania to cover the annual Groundhog Day ceremony held there every February 2nd. As the events of the day unfold, Murray's character reveals himself to be an egocentric jerk.

At the end of the day, Murray and the TV crew are trapped in Punxsutawney by a snowstorm and have to spend the night.

The next morning when Murray's character wakes up, he finds that it is not February 3rd, it is February 2nd all over again. Everything happens just as it did in what Murray's character perceives as the previous day, while all the other characters perceive that everything is happening for the first time.

Murray reacts in the same self-centered, egocentric way, only to find that he is caught in a kind of continuous loop as Groundhog Day repeats time after time in what seems to be an ongoing, endless cycle. Gradually, however, Murray's reactions to events change. Over a period that seems an eternity of Groundhog Days, his actions evolve to a point at which he deals with events in a caring and helpful manner. The morning after he has dealt with all of them in what would seem the best possible way, he wakes up to find that it is finally February 3rd.

He is now able to move on.

I see this as an allegory for life on Earth because I believe that we incarnate time after time and face similar challenges and obstacles until we finally learn to deal with them in the best ways possible.

Let me relate one last story. In it, a young man receives a call to adventure, but he refuses the call. It involves Jesus and is found in all three synoptic gospels. This account is from Mark 10:17-23, the New International Version (NIV) translation:

As Jesus started on his way, a man ran up to him and fell on his knees before him. "Good teacher," he asked, "what must I do to inherit Eternal Life?"

"Why do you call me good?" Jesus answered. "No one is good—except God alone. You know the commandments: 'Do not murder, do not commit adultery, do not steal, do not give false testimony, do not defraud, honor your father and mother.'"

"Teacher," he declared, "all these I have kept since I was a boy."

Jesus looked at him and loved him. "One thing you lack," he said. "Go, sell everything you have and give to the poor, and you will have treasure in heaven. Then come, follow me."

At this the man's face fell. He went away sad, because he had great wealth.

Jesus looked around and said to his disciples, "How hard it is for the rich to enter the Kingdom of God!"

The disciples were amazed at his words. But Jesus said again, "Children, how hard it is to enter the Kingdom of God! It is easier for a camel to go through the eye of a needle than for a rich man to enter the Kingdom of God."

Well, there you are. If ever a person received the call to adventure, it was this rich man. If he answers the call, he will be on the road to Eternal Life, which I believe is a kind of merger with the Infinite Mind and what Jesus also referred to as entering the Kingdom of Heaven. In other words, as with Minos and Daphne, the promise is that if the young man answers the call, he will develop and eventually experience the ecstasy of a relationship with the divine. But first, as was the case with those two, he must give up his earthly treasure and put his ego aside, just as Bill Murray's character finally did in *Groundhog Day.* However, as was the case with Minos and Daphne, as well as many of us today, he was much too attached to his earthly wealth and egocentric pride to do so.

So, returning to my personal story, there I was, tired of the ad game and ready to move to a higher plane, but held in place by golden handcuffs like the rich young man in the story above. I was ripe for the call, and naturally it came. It was not easy to turn away from that earthly treasure, but I did. I started writing. And I loved it.

Like any hero's adventure, it was frightening to take that first step, to answer the call, and the adventure became even more frightening as it continued. It wasn't long before a good deal more money was going out than coming in. I started doing freelance work to stem the flow, but nevertheless, I had to dip into savings in a big way before things began evening out. What I went through was no fun, but let it be sufficient to say I had to fight my own dragons and demons and to confront the fears that told me I ought to get my nose pressed back against the grindstone of the workaday world and get a real job. But, as in any hero's adventure, when the going got really, really tough, unseen hands, the support of the divine, stepped in. But that story, my friends, will have to wait for another book.

I hope that, as others who have taken such a journey, I have finally arrived back at the beginning. It is true that, like the heroes of old, I'm on a higher plane of understanding, mentally and spiritually, but sad to say, I am not better off financially. Nevertheless, I feel more wealthy in a number of ways than I did before I answered the call, and I know that as long as I am pursuing the life I came here to live, invisible hands will clear a path for me and ensure I have enough to eat and a roof over my head. I've entered the Kingdom—am now a loyal subject and reap the benefits. I am no longer afraid of poverty or death. In place of fear and doubt, I sense the presence of awesome power ready to help whenever the going gets tough.

Chapter Five: More Implications of the New Theory

The Source, aka, Infinite Mind fosters growth and life, and so perhaps it should not come as a surprise that prayer appears to double the success rate of in vitro fertilization procedures that lead to pregnancy, according to a study published in the September, 2001 issue of the *Journal of Reproductive Medicine*. The findings reveal that a group of women who had people praying for them had a 50 percent pregnancy rate compared to a 26 percent rate in the group of women who did not have people praying for them. In the study, led by Rogerio Lobo, chairman of obstetrics and gynecology at Columbia University's College of Physicians & Surgeons, none of the women undergoing the IVF procedures knew about the prayers on their behalf. Nor did their doctors. In fact, the 199 women were in Cha General Hospital in Seoul, Korea, thousands of miles from those praying for them in the U.S., Canada and Australia. According to Dr. Lobo, "The results were so highly significant they weren't even borderline. We spent time deciding if it was even publishable because we couldn't explain it."

This is not the only study to indicate that prayer can have a significant effect on matters of health. Another example comes from Randolph Byrd, a cardiologist, who over a ten-month period used a computer to assign 393 patients admitted to the coronary care unit at San Francisco General Hospital either to a group that was prayed for by home prayer groups (192 patients), or to a group that was not prayed for (201). This was a double blind test. Neither the patients, doctors, nor the nurses knew which group a patient was in. Roman Catholic as well as Protestant groups around the country were given the patients' names, and some information about their conditions. The various groups were not told how to pray, but simply were asked to do so every day.

The patients who were remembered in prayer had remarkably different and better experiences than the others. They were three times less likely to develop pulmonary edema, a condition in which the lungs fill with fluid; they were five times less likely to require antibiotics. None required endotracheal intubation (an artificial airway inserted in the throat), which twelve in the un-prayed-for group required. Also, fewer prayed-for patients died, although the difference between groups was not large enough to be considered statistically significant.

A third study indicating that prayer may have positive health effects is at this writing scheduled to be published in the *International Journal for Psychiatry in Medicine*. A team from the University of California at Berkeley found that Christians and Jews who regularly attended services lived longer and were less likely to die from circulatory, digestive and respiratory diseases. Devotees of Eastern religions were not surveyed. The study examined links between religious attendance and cause-specific mortality ovef a thirty year period in 6,545 residents of Alameda County, California. Even after adjusting for variables like health and frequency of exercise, religious devotees lived longer without succumbing to disease.

"At this point it's a puzzle why there should be this pattern," said the study's author, Doug Oman, Ph.D., a lecturer at Berkeley's School of Public Health. "It's likely a stress-buffering resource. Regular attendance at services can give people an inner peace that is unshakable. That results in less wear and tear on their bodies."

It is not surprising that Dr. Lobo of Columbia University and Dr. Oman of Berkeley are puzzled by the results of their own studies. These men are card-carrying members of the religion of materialistic science, and this religion still operates under the misconception that the body arises spontaneously out of the mother's egg, the father's sperm and the genes of both parents, and that the body's brain gives rise to thought. They believe that awareness and thought are the result of electrons jumping across synapses and that thought remains inside the skull. We, of course, know why and how prayer works. It does so because all things, including people and their bodies, are products of the universal subconscious mind. People's individual subconscious minds are diligent in their efforts to create what the owners' conscious minds believe, and subconscious minds are part of the universal subconscious mind. The belief of those praying that their prayers will be answered is impressed upon the subconscious, and the subconscious faithfully acts upon the bodies of those being prayed for.

An organization exists that has as its purpose the study of what prayer techniques produce the best results. It was founded by Christian Science practitioners who have been at this since 1975. (The name of the organization is Spindrift, Inc., and the address is: P. O. Box 452471, Ft. Lauderdale, Florida 33345.) Resting next to my keyboard at this moment is a document an inch thick, printed on both sides of standard letter-size paper called, "The Spindrift Papers." It gives detailed information of prayer experiments conducted under rigorously controlled conditions.

The first question Spindrift researchers sought to answer is, does prayer work? The answer as we already know, is yes. In one test, rye seeds were split into groupings of equal number and placed in a shallow container on a soil-like substance called vermiculite. (For city dwellers, this is commonly used by gardeners.) A string was drawn across the middle to indicate that the seeds were divided into side A and side B. Side A was prayed for. Side B was not. A statistically greater number of rye shoots emerged from side A than from side B. Variations of this experiment were devised and conducted, but not until this one was repeated by many different Christian Science prayer practitioners with consistent results.

Next, salt was added to the water supply. Different batches of rye seeds received doses of salt ranging from one teaspoon per eight cups of water to four teaspoons per eight cups. Doses were stepped up in increments of one-half teaspoon per batch.

A total of 2.3 percent more seeds sprouted on the prayed-for side of the first batch (one teaspoon per half-gallon of water) than on the unprayed-for side (800 "prayed-for" seeds sprouted out of 2,000, versus 778 sprouts out of 2000 in the not-prayed-for side). As the dosage of salt was increased the total number of seeds sprouting decreased, but the number of seeds which sprouted on the prayed-for sides compared to the unprayed-for sides increased in proportion to the salt (i.e., stress). In the 1.5 teaspoon batch, the increase was 3.3 percent. In the 2.0 teaspoon batch, 13.8 percent. In the 2.5 batch, 16.5 percent. In the 3.0, 30.8 percent. Five times as many prayed-for seeds in the 3.5 batch sprouted (although the total number which sprouted was small as can be seen from the chart below). Finally, no seeds sprouted in the 4.0 teaspoon per eight cup batch.

What this says is what people in foxholes with bombs going off around them have always known: the more dire the situation, the more helpful prayer will be. Up to a point. There comes a time when things are so bad that nothing helps.

This experiment was also conducted using mung beans. The solution of salt and water ranged from 7.5 teaspoons per half-gallon of water to 30.0 teaspoons per half-gallon. The increase in the number of sprouts for the prayed-for side ranged from 3.3 percent to 54.2 percent.

Salt	Control	Grown	Prayed for	Grown	% Increase
1.0	2,000	778	2,000	800	2.3
1.5	3,000	302	3,000	312	3.3
2.0	3,000	217	3,000	247	13.8
2.5	3,000	454	3,000	528	16.3
3.0	3,000	52	3,000	68	30.8
3.5	3,000	2	3,000	10	400.0
4.0	3,000	0	3,000	0	0.0

Next an experiment was constructed to determine whether the amount of prayer makes a difference. This involved soy beans in four containers. One container was marked "control" and not prayed for. The other three were marked X, Y, and Z. In each run of the experiment, the X and Y containers were prayed for as a unit, and the Y and

Z containers as a unit. So, Y received twice as much prayer as either X or Z. The Y container also had twice as many soybeans germinate. The results were in proportion to the amount of prayer.

This is reminiscent of a principle set forth by Napoleon Hill in his perennial bestseller, Think and Grow Rich. He wrote this granddaddy of all self-help books in the late 1930s and updated it in 1960. One chapter is devoted to the principle he called "The Master Mind." Hill suggested that whatever project or purpose or goal an individual had, it could be advanced and achieved most readily by bringing together a group of people to apply their unified brain power to it. Hill never used the word prayer nor did he suggest people sit around and pray. But he did liken a group of minds at work on a project to a group of storage batteries connected together in a series to produce much more power than any single battery possibly could on its own. He wrote, "When a group of individual brains are coordinated and function in harmony, the increased energy created through that alliance becomes available to every individual brain in the group." He cited several examples, including the remarkable successes of Henry Ford and Andrew Carnegie, each of whom had a group of colleagues around him working and pulling together on common goals. Hill wasn't referring only to innovative thinking that leads to marketing and sales results. He was talking about much more, of creating an aura that leads to favorable events taking place, or to what might be considered by Materialists as "good breaks." The Master Mind creates a force with a life of its own, a force I've called grace later in this chapter. This is the force that works much like unseen hands. A case Hill cited was that of Mahatma Gandhi, who led the successful non-violent revolution that freed India from British Colonial rule. Hill wrote, "He came to power through inducing over two hundred million people to coordinate, with mind and body, in a spirit of harmony, for a definite purpose."

If one person can make 54.2 percent more saltwater-soaked mung beans sprout with his mind, imagine what two hundred million can do. They toppled a government which had been in power for more than 150 years, and they did it without firing a shot.

The power of the Master Mind is another good reason to become part of some sort of spiritual brotherhood. You may want to organize a study group as well. I propose that groups range in size from three to ten and that they be dedicated to the expanded consciousness and spiritual growth of each of its members. A group ought to meet a minimum of twice a month, or more frequently if possible. For a period of about five years, I participated in three such groups and experienced quantum leaps in my own development.

The group ought to study a text such as this, or the Bible, or the Bhagavad-Gita. Share your thoughts, your individual interpretations of what you study, and ideas about how to put what you've learned to use. You'll also want to set aside a time at each meeting to share personal concerns, fears and troubles. Prayer is one of the tools your group can and ought to employ.

For example, you may wish to pray for someone who is trying to get pregnant. As the study cited at the beginning of this chapter demonstrated, prayer can be particularly effective in this regard. This makes perfect sense, since the Life Force's goal is to foster life. In the Columbia study, the people praying were from Christian denominations and were separated into three groups. One group received pictures of the women and prayed for an increase in their pregnancy rate. Another group prayed to improve the effectiveness of the first group. A third group prayed for the two other groups. According to the authors of the study, anecdotal evidence from other prayer research has found this method to be most effective.

The three groups began to pray within five days of the initial hormone treatment that stimulates egg development, then continued to pray for three weeks.

Besides finding a higher pregnancy rate among the women who had a group praying for them, the researchers found that older women seemed to benefit more from prayer. For women between 30 and 39, the pregnancy rate for the prayer group was 51 percent, compared with 23 percent for the non-prayer group. This would seem to parallel the Spindrift study in that those who needed help the most, up to a point at least, saw the biggest gains from prayer versus no prayer. With Spindrift, it was salt-soaked rye and mung beans. With the Columbia study, it was older women.

Is there anything more the Spindrift researchers learned which would be helpful to know?

The quality of prayer is a factor in how effective it is, as is the quantity or amount of prayer. Like anything, practice makes perfect. More experienced practitioners got better results than less experienced practitioners. Get in the habit of praying. Do not save the practice of it only for the times bullets are flying overhead, or the airplane you're in goes into a tailspin.

The Spindrift research also gives us clues on how to pray. First, you need to know what you're praying for. Some experiments were conducted in which the prayer practitioner was kept in the dark about the nature of the seeds he was praying for. He or she did not know what kind of seeds they were or to what extent they had been stressed. Results showed a drastic reduction in the effect. The researchers concluded that the more the person praying knows about that which is being prayed for, the greater the positive effect of the prayers.

Another experiment measured the efficacy of "directed" versus "non directed" prayer. Directed prayer was that in which the practitioner had a specific goal, image, or outcome in mind. He attempted to steer the seeds in a particular direction. A parallel in healing might be for blood clots to dissolve or for cancer to isolate itself in a particular place in the body where it could be cut out. In the seed germination experiments it was praying for a more rapid germination rate. Non directed prayer used an open-ended approach in which no specific outcome was held in the imagination. The person praying did not attempt to imagine or project a specific result but rather to ask for whatever was best for the seeds in an open-ended spirit of "Thy will be done." Both approaches worked, but the non-directed approach appeared to be more effective, in some cases producing twice the results.

Using the non-directed approach is bound to conflict with the beliefs of many who hold that one must visualize a specific result and hold it in his mind. No doubt in some cases this works. The problem is that we humans often do not know what the best outcome of any given situation might be. The theory put forth by Spindrift researchers is that prayer reinforces the tendency of an out-of-balance organism to return to balance. It enhances the Life Force, which you know by now is the opposite of entropy. In other words, the goals of nature are harmony and growth, and prayer supports this. To quote the Spindrift research document, "If the power of holy prayer does, indeed, heal, then such a power will be manifest as movement of a system toward its norms since healing can be defined as movement toward the optimal or 'best' conditions of form and function." The Spindrift researchers did not try experiments in which prayer was used to try to prevent seeds from germinating. If they had, and if what they say here is true, this would not have worked.

Chances are that we don't know the best way for an organism to achieve balance. Likewise, the subjective mind and the Big Dreamer push in the direction of growth and evolution. The best outcome of a situation will have growth of some kind as a result. In this way, nature achieves harmony and balance, or the healing of a splintered soul.

This may be bad news for anyone who picked up this book thinking it would provide a formula for conjuring riches without the conjurer having first to change. If a new Mercedes will not help foster your spiritual development or someone else's, you're wasting your time praying for it no matter how clearly you can picture in your mind a shiny new one appearing in your driveway.

Here's what Jesus' brother James had to say about this: "You do not have, because you do not ask God. When you ask, you do not receive, because you ask with wrong motives, that you may spend what you get on your pleasures." (James 4:2-3.) Our motives need to be in line with the goals of the universe. God is not in the business of satisfying our selfish whims. He wants something much more valuable. He wants us to evolve.

Finally, here is what Thomas Troward has to say about creating mentally, or prayer. The text below is taken from the modern English version of his *Edinburgh Lectures on Mental Science* as they appear in *How to Master Life,* first published in 2002 by The Oaklea Press:

> *Some people possess the power of visualization, or making mental pictures of things, to a greater degree than others. This faculty may be employed advantageously to facilitate the working of the Law. But those who do not possess this faculty in any marked degree, need not be discouraged by their lack of it. Visualization is not the only way to put the law to work on the invisible plane. . . .*
>
> *We must (simply) regard our mental creations as spiritual realities and then implicitly trust the Laws of Growth to do the rest.*

Our minds are in touch with the Big Dreamer, which has access to and is immersed in all the information needed in any situation. The way to the best result may be exactly the opposite of what we expect, which means that it is always smart to put things in God's hands and pray for the best possible outcome. Consider this outcome already accomplished. Do not attempt to explain to the universal subconscious what course it should take to arrive at the desired destination. Let the Infinite Mind find the best way.

The readings of Edgar Cayce have something to say about all this. Suppose, for example, someone you love is suffering from an addiction such as alcoholism. You want to help, but how? Prayer is certainly one action you can take. But if this loved one is not yet interested in changing, what form of prayer is best? Cayce described two kinds: direct and protective. In direct prayer, you ask for a specific healing to take place. Such a prayer is appropriate, he said, only if the target of your prayer has asked for such efforts and desires that outcome. In this case, you are adding energy to a process of change that he or she has already willed. If the person you are praying for is not in sympathy with your efforts, your prayers may actually aggravate the problem. In such a case, protective prayer is best. With this type of prayer, ask that the person be surrounded and protected by the forces of love and healing, while at the same time allowing that individual his or her own free will in choosing whether or not to change.

You recall my prayer to have events take place that would wipe out the bad karma I felt was hanging over me. This was a direct prayer, and I got just what I asked for— although it was perhaps more than I bargained for. Nevertheless, the experience caused me to grow. Even my wife divorcing me turned out for the best. I suspect that neither of us had been happy for quite some time. Today, I am remarried and have children that are delights of my life. Had I remained married to my first wife, they would never have been born.

I've often seen direct prayers answered. They always seem to result in growth, and sometimes in unexpected ways. A friend in one of my study groups, for example, recounted a story recently about how we should be careful what we ask for because we just might get it. Her son had spent a miserable autumn and winter, first on the bench of his high school football team and then on the bench of the basketball squad. Lacrosse season was getting underway and the first game was scheduled for that afternoon. Her son was in the starting lineup. At last he would have a chance to show his prowess. "Please, Lord, have him score a lot of goals," she prayed. "Let him be the star of the team today."

She was thrilled and amazed as she watched the game. Not only did his team win, her son scored all the goals for his side. He was all over the place; seemed to be everywhere at once. She patted herself on the back and praised the Lord all the way home.

But her joy was short-lived. When her son came home he was depressed.

"What's wrong?" she asked. "You should be feeling good. You were sensational."

"Aw, Mom. No I wasn't. I was a ball hog. It was like I never gave anyone else on my team a chance. It didn't even feel like it was me out there playing. It was as though I was possessed, or something. Like someone else scored those goals—not me."

My friend had wanted a feel-good experience for her son and herself. What was received was a growth experience for her son, and a learning experience for herself.

Next time you pray, think about what you are really asking. Will it help you or someone else grow? Think, too, about what is happening in the unseen world as a result. Here is what Betty Eadie experienced when out of her body while clinically dead. In *Embraced by the Light* she wrote:

> *I saw many lights shooting up from the earth like beacons. Some were very broad and charged into heaven like broad laser beams. Others resembled the illumination of small penlights, and some were mere sparks. I was surprised as I was told that these beams of power were the prayers of people on earth. I saw angels rushing to answer the prayers. They were organized to give as much help as possible. As they worked within this organization, they literally flew from person to person, from prayer to prayer, and were filled with love and joy by their work.*

I imagine what Betty saw was a metaphor constructed by her mind. It seems logical to me that mental constructions are how we experience the spirit realm. Heaven and hell are what we imagine and believe them to be. Nonetheless, her vision is one of beauty.

Let's sum up what we need to keep in mind about prayer:

- Belief is the key. Believe that what you pray for already exists in the realm of spirit and that it is only a matter of time before it manifests on the physical plane.

- Practice makes perfect, or in other words, experienced prayer practitioners receive the best results.

- Quantity is a factor. More minds at work praying brings more and better results.

- The more a person or group knows about the subject of their prayers, the better.

- If the desired outcome is clear, visualize it, and pray for it. Consider it an accomplished fact. But do not tell the universal subconscious how to arrive at this outcome. Let it find the way.

- If the best outcome is not clear, prayers should be kept general in nature. Pray for the best outcome. The universal subconscious knows.

- For good results, the purposes of Infinite Mind need to be served by our prayers. This includes spiritual growth and development, the healing of the soul and life, or in the case of physical healing, the bringing of a stressed body or physical system into harmony or balance.

A word of caution may be in order. Once you ask in prayer for something, be prepared to experience events you might never have chosen for yourself. At first it may seem like a roller coaster ride, and you may wish you could get off. What seems to be a disaster may happen. You may get transferred to another city by your employer. Your wife may ask for a divorce. Your apartment building might burn down. Of course, you might get a promotion, win the lottery, be offered a dream job out of the blue. But I predict that whatever happens, the path won't be easy. Growth takes effort and means change. A transformation must take place in you.

Most people resist change. This is their ego fighting for what it thinks is survival. Even if you want to transform, it won't be easy. So you might as well anticipate a number of character-building challenges. A friend in one of my study groups relates a story which illustrates my point. One day, feeling frustrated with a situation he had to deal with, he prayed, "Lord, give me more patience." Soon, he found himself in a situation that took every ounce he could muster.

Later, it came to my friend that we have to learn by doing—that "practice makes perfect." He was sent what he needed—an opportunity to exercise patience. He'd thought God would hand him more patience as a gift but more patience came to him in the only way it could—he had to learn and eventually earn it.

When you ask for change, don't be surprised if some of the change that occurs is unrelated to the central issue. For example, when I was making the transition from advertising agency president to writer and publisher, my car started falling apart. It wasn't a particularly old car, but nevertheless, one thing after another went wrong. I believe now this was an outward sign of inward change taking place in my life. Since everyone and everything is connected, and part of the whole, I guess we need to expect this sort of thing. Trust. Continue asking for guidance. If you think you have an answer but aren't positive, don't do anything precipitous. Ask for further guidance, confirmation, or some form of reassurance.

And expect to be helped by the invisible hands of grace. What is grace? Grace is what happens when the universal subconscious is working in people's lives to insure or further growth and development. To the untrained eye, grace appears to be a set of mysterious or unexplainable conditions, events and phenomena that support, nurture, protect or enhance human life and spiritual growth. Grace works in all sorts of ways. The forms of grace seem to be universal. Our immune systems, for example, are tied to it. Modern medicine has only a vague idea why one person exposed to an infectious disease will come down with it and another experiencing the same level of exposure will not. On any given day, in practically every public environment, potentially lethal microbes and viruses on surfaces or floating in the air are too numerous to estimate. Yet, most people do not get sick. Why? Doctors would say it is because most people's resistance is fairly high. But what do they really mean? That most people are not rundown or depressed? Perhaps. Not everyone who is rundown and depressed contracts an infectious disease. Yet many do who are perfectly healthy and in good shape.

In some cases, however, getting sick may be an act of grace. At one point after leaving the ad business, I got discouraged. I very nearly threw in the towel with respect to my dream of writing books. I concluded it was time to return to the rat race and took several steps in that direction. I actually had promotional materials printed and was putting together a mailing list. It wasn't what I wanted to do, but I was worried that I'd never make it as a writer. I was afraid my money would run out. So I stood on the brakes and was in the process of making a U-turn. Then grace stepped in. I got sick. I caught the flu. It was a bad case that lasted almost two weeks, and it gave me plenty of time to think.

Whenever I get sick I always ask myself why. Sometimes the answer is that I'm pushing myself too hard and need to slow down. This time, my system was telling me

I'd be making a big mistake to reverse course. I was as certain of the message then as I am of it now. For me, the
viola twanged its low-pitched note.

Most people do not get sick and face death because
that is not necessary for their personal evolution or the
evolution of humankind as a whole. In my case, catching
the flu was just what needed to happen. It was my wake
up call, and it worked. I decided to stay the course.

I'm sure that sort of thing happens every day. I didn't
need a really big, life-threatening illness to get my attention because I'm on the lookout for such things. But think
about those who have stalled in terms of personal growth
and have no idea how these things work. A serious illness
can be the wake-up call needed to snap them out of it and
get them back on track. ("My God, if I die I won't be able
to accomplish X, Y, and Z, and I really want to do that.
Please, God, let me have another chance!")

Or, it may be that nothing is going to do the trick.
They've truly reached a dead end. This individual's higher
self and the Big Dreamer may have come to the conclusion
that the usefulness of the present incarnation has come to
an end and that it's time for the individual to move on.
This person will not pull through. Those who are left behind may grieve and wonder why, but this course will allow
the person's soul to assimilate what has been gained in this
lifetime. The individual will now be free to be born again
into a new body and begin a new life cycle of growth. In

the big scheme of things, getting on with a new incarnation and moving ahead will be preferable to figuratively treading water until the individual has lived out his or her allotted three score years and ten.

By the way, the grace of resistance is not limited to infectious disease. Have a chat with a state trooper who has been on the scene of a number of motor vehicle accidents. Ask the trooper what percentage of crashes appeared fatal when he first arrived, and how many actually turned out to be. You're likely to hear some amazing stories of cars or trucks smashed beyond recognition, metal so collapsed, twisted or squashed the trooper will say, "I don't see how anyone could have survived. And yet the person walked away without a scratch," or with only minor injuries. How is it scientifically possible for metal to collapse in such a way as to conform perfectly to the shape of the human body contained inside? Nevertheless, I'm willing to bet the trooper will tell you that this happens more often than not.

When she was about a year old, my now twenty-four-year-old daughter body-surfed down the steep flight of stairs from our kitchen to our basement—not just once, but twice. Another time, a babysitter turned her back while changing a diaper, and the same daughter rolled off the counter top and fell straight to the bare kitchen floor. Any of these three falls could easily have been fatal. None caused so much as a bruise.

Almost everyone has experienced a close call that could have killed him. In an earlier chapter I promised I would tell a story I believe to have come about through divine intervention. It happened when I was fourteen. I darted across Jefferson Davis Highway without looking properly. At the time, it was the main north-south highway on the East Coast. This particular stretch had six lanes (three north and three south) with a grass median. A car struck me in mid-stride. Maybe it was the way the car bumper caught my foot that lifted me into the air, but I should have been pushed down and run over. Instead, I was lifted up, seemed to fly through the air, and landed on the grass median. The driver was certain I was dead—until I stood up and dusted myself off. I didn't have a scratch. The only evidence of the accident was the stain on my trousers where I'd slid on the grass as I came to a stop. Also, both my shoes were missing. I found them eighty or a hundred feet away where the car had screeched to a halt. If the laws of Newtonian physics had been working that day, I wouldn't be here to put this down on paper.

How, physically, was I lifted into the air? Were angels responsible? Did they pierce the veil, reach out and lift me up? It certainly seems the only way what happened could have happened. But whatever the case may be, the phenomenon called grace came to my aid, and I lived to be an adult. As a result, I grew and studied and learned enough to enable me to write this book. And you know what? If this book didn't exist, at least some who would have read it, perhaps including you, would die before it was necessary for them to do so. They'd check out of this life before accomplishing the objectives set out before they were born. So the angels that lifted me up did them a big favor as well as me.

The role of grace is to help advance the evolution of souls. This fits with the forms of communication with your higher self or soul discussed in the last chapter: The voice telling you to wake up and take note because the mother of your children is approaching, the clairvoyant message that says someone you love is in trouble, a dream that brings the answer to a question. Or grace may manifest as the answer to a prayer.

If you want to take advantage of the higher intelligence available to you, it makes sense to align yourself with grace. Make the evolution of your soul, and the souls of others, your number-one and number-two priorities respectively. The forces of the universe will fall in behind you to help make this happen. I know, because I experience this daily.

Let me give another example. Twenty-five years ago, a friend in one of my study groups and his wife quit their full time jobs in order to attend seminary together. They both had to work part-time and even then were only able to bring in enough to just get by. Unexpected bills arrived, as they always do. In this case, they totaled $578, money they simply didn't have. The couple's bank balance registered zero. They had no place to turn. Creditors were calling. Our group prayed that the money they needed would come to them. My friend and his wife prayed, too.

Two days later, the couple received an envelope in the mail from the IRS saying that their petition had been reviewed. Their tax return from two years prior had been found to be in error. Along with the notice was a check made out to them for $588.

Good timing? True. But the amazing thing was, the couple had not filed a petition. Nor had they filed an amended return. Somehow, the IRS had done this recalculation on its own. The couple rummaged in their files and pulled out their return from two years prior. The IRS was correct. They found the error that had been referenced.

In my experience, the IRS is not in the mode of helping people out this way. It was grace that brought them that check because they needed the money to stay in school. Seminary was helping them grow, and their growth and the degrees they eventually received would someday allow them to help others grow.

Why was the check for ten dollars more than they needed? Maybe, since our group met at a restaurant over breakfast on Thursday mornings, grace wanted to pick up the tab.

You may also be familiar with the story of psychiatric pioneer Carl Jung that he related in an article called "On Synchronicity." Jung had a patient, a young woman, who was the type who thought she knew everything. She was well educated and used highly polished rationalism as a weapon to defend herself against Jung's attempts to give her a deeper, spirit-based understanding of reality. Jung was at a loss as to how to proceed and found himself hoping something unexpected and irrational would happen in order to burst the intellectual bubble she'd sealed herself within.

One day, they were in his office. He had his back to the window, and she was talking. She was telling him about a dream she'd had the night before in which she'd been given a golden scarab—an expensive piece of jewelry. Jung heard something behind him tapping at the window. He turned and saw a large flying insect knocking against the pane on the outside, trying to get in. He opened the window and caught the insect. It was a scarabaeid beetle, or rose-chafer *(Cetonia aurata),* whose golden-green color resembles a gold scarab.

Jung handed the beetle to his patient with the words, "Here is your scarab." This poked the desired hole in her rationalism, exactly as Jung had hoped. She had dreamed about the gift of a scarab and now it had happened. What she received, however, was a much bigger gift. She was shown through grace that all is of one mind, that she was a single character in the larger dream of life. Grace worked, as always, to advance spiritual development.

Perhaps you are now saying to yourself, these sorts of things never happen to me. This Martin fellow is living in a fantasy world. To this I will ask, are you making an effort to advance and grow spiritually? If so, are you on the lookout for acts of grace? Do you expect synchronicities? You must be open to them and permit them to happen. You must expect them. If Jung hadn't been looking for an unusual, irrational occurrence, if he hadn't been expecting it, he might not have bothered to open the window. And if I didn't expect to find the quotation I need or the answer to a question when I walk into a library or bookstore, I doubt I ever would. If I did happen to find what I need, I'd chalk it up to coincidence, wouldn't I? I'd tell myself it probably wouldn't happen again. And I probably wouldn't see it when it did. As I've said a number of times in this book, you usually get, or don't get, what you expect.

Expect grace to happen. Be on the lookout and it will.

Let's suppose you have decided to strike out on the path of spiritual growth. One way to insure you'll be helped along by grace is to cut off avenues of retreat. This is what the couple in seminary had done. Both had quit their jobs. It seems possible they received help partly because there was no other alternative. I'm reminded of the general of ancient times who took his army across a sea to fight a distant enemy. As soon as he and his men landed, he ordered the ships that had brought them burned, cutting off all means of retreat. This created a big incentive—his men had no choice but to win or die. Of course, they won. But the question is, was it purely the will to live that led to victory? Isn't it also possible they got some breaks because of the desperate situation they were in? If you cut off all means of retreat, your subconscious mind, or perhaps your guides or guardian angels will be left with no other alternative but to help you. I believe this is precisely what they want to do. They want you to make progress. They want you to wake up and live. If they do allow you to stumble and fall, it will be because this is necessary before you can climb to new heights.

Making the effort to grow spiritually is difficult, and it takes courage. Perhaps most difficult is being totally honest with yourself about yourself and your surroundings. You may not like what you see, and this can be painful. Just keep in mind that growth means change. Your goal is to become the best you can be. The problem is that most of us think we are just fine the way we are. Our ego self does not want to change for fear that change will wipe it out of existence. Of course, that isn't true. Your sense of self will never disappear, and the new you will be happier, stronger, more vital, alive, awake and aware. But this transformation will take effort. There will be hardships to overcome. Pursuing spiritual growth is not an undertaking for the faint of heart or the lazy.

It's human nature to be lazy. Everyone is lazy part some of the time. Unfortunately, a lot of us are lazy a lot of the time. We think the world owes us a living. We think we can get by without trying all that hard. What we need to understand is that life will be difficult whether we choose to stay put and "play it safe," or instead strike out on the wondrous adventure of growth.

Getting on the right path is what each of us needs to do, and what I hope for you. Remember. You exist to evolve, and this means going the direction of the flow. Your subconscious mind and the Big Dreamer want you to join them. They want a relationship that will eventually lead to a merger. But don't kid yourself, you will have to make major adjustments in your life, and these take strength and courage. But as you get to know your true, innermost higher self, and form a relationship, you will grow to the point where you would not trade the relationship or the growth for anything.

* * *

Seven steps to finding and taking your path:

1. Recognize that a force exists that's not yet acknowledged by science. The opposite of entropy, it fosters evolution and growth.

2. Form an alliance with it. Ask for direction and guidance.

3. Expect a "call to adventure," which is the Life Force beckoning for you to follow your path to a higher level of understanding.

4. Look for communication and guidance, which will come through intuition, the Scriptures, the written word, others, and your circumstances.

5. After answering the call, expect a crisis. Resist the temptation to abandon the quest. Press ahead. Watch grace come to your aid.

6. Major adjustments in your life will probably be required. Make them.

7. Self-actualization will come as you obey your inner voice and accomplish the work assigned. Abundance, joy, and fulfillment will be yours.

Chapter Six: Why Are You Here?

*Before it incarnates, each soul enters into a sacred contract
with the Universe to accomplish certain things. It enters
into this commitment in the fullness of its being. Whatever
the task that your soul has agreed to, all of the experiences
of your life serve to awaken within you the memory of
that contract, and to prepare you to fulfill it.*

—Gary Zukav
The Seat of the Soul

Did we incarnate and enter the physical realm with a
specific mission or missions to accomplish as stated by
Gary Zukav in the quotation above? Many believe so. If
so, some aspect of ourselves would have to have existed
before we were born. Let's examine this.

To the Scientific Materialist each individual human is
an assembly of parts, the same as my twenty year old Toyota Land Cruiser. We are built of a brain, heart, blood ves-
sels and muscles. The Toyota has fuel injectors, pistons, a
crankshaft and valves. It was put together in a factory. We
were assembled in the womb. Fortunately for the Land
Cruiser, I've taken care of it so it has outlasted many of

its contemporaries and enjoyed a relatively good existence for an SUV.

Humans, too, seem to be subject to the same whims of fate. One may be lucky and be born to a wealthy, well-educated family in the West or unlucky and enter the world in Somalia, Afghanistan, or some other Third World nation where living conditions are miserable and opportunities for a good life are largely nonexistent.

Or is it luck? We know from our review of the facts that a human being is not an assembly of parts. Rather, behind, supporting, and giving life to the physical body is a thought construction—an assembly of memories that exist in the medium of mind or spirit. In his book, *A New Science of Life: The Hypothesis of Morphic Resonance* (Park Street Press, 1995), the British biochemist, Rupert Sheldrake, presented a theory that seems plausible based on the New Cosmolgoy. He wrote that our physical bodies are projections of our personal morphogenetic field that combines with the fields of our parents and other members of our species. The resulting morphogenetic field is what shapes our body in the womb, while genes produce and release the necessary proteins at the appropriate times.

Our bodies are not made of separate pieces. Each is a unified whole. The human body's abilities were shaped by the evolution and experiences of our species in an unbroken chain dating to the first life on earth. We are each separate only to the extent that we identify ourselves as such. At a deep level, we are all totally at one with the Source— the universal mind from which we evolved. Indeed, as we each grow spiritually and move through the stages described earlier, we will come to sense our connection to all of creation. If we answer the call to adventure when it comes, we will experience growth that will lead to a new way of seeing the world. This shift may be accompanied by a sudden insight—an epiphany.

As an individual who has had this insight grows in wisdom, experiences life, and moves into and through middle age, the idea of fate as whimsical and arbitrary will increasingly seem contrary to personal experience. The Nineteenth Century German philosopher Arthur Schopenhauer, for example, observed in one of his essays that when an individual reaches an advanced age and looks back over his or her lifetime, the lifetime will seem to have followed a consistent plan as though composed by a master storyteller or novelist. Specific events and meeting of individuals that seemed at the time to have come about by chance turn out to have been essential components in a constant storyline.

If this is so, and my personal experience says it is, we are compelled to ask who wrote the story?

Today, Schopenhauer would have said that it was an individual's subconscious mind. He would note that our dreams are composed by part of us of which we are unaware. He'd argue that our whole life is created by a subconscious aspect of ourselves that he labeled the "will within." This will within merges with those of others so that the whole of human existence comes together like a symphony.

As we now know, according to the New Cosmology, only one organism with one single mind exists, and this mind is having the dream we call reality. We have trouble understanding this because we are each an aspect of the organism—you, me, and everyone—are each characters in the dream, and we each possess a unique point of view. Our individual perspectives are from where we happen to sit in relation to the whole. This point of view constitutes our personal reality—the environment our minds unconsciously create. Yet our individual subconscious minds or souls are part and parcel of the whole, which is the Infinite Mind that is the Big Dreamer.

This idea is not new. More than 400 years ago John Donne wrote, "No man is an island." Our lives are intertwined. As a piece of the continent of mankind, we have roles to play that affect other parts of the mainland. An individual we meet apparently by chance becomes a key player in the story of our life, just as in turn we play key roles in the lives of others whether or not we realize it.

What is the part of us, our personal puppeteer with its unique point of view, that compels us to play our different roles at different times in the giant dream of humanity?

And why aren't we aware of it? Was this part created at the moment the egg, the sperm and the morphogenetic fields of our mother and father united with our own unique morphic field? Did it develop as our egos developed, a sort of parallel construction? We've said that an unconscious part of our conscious mind exists that stores the memories and programming of this life. The other part of ourselves that is beyond our conscious awareness is our subconscious mind or soul. This is our personal morphic field built up through our own evolution that began when life began. This part is our puppeteer. It is privy to the big dream. It is a player in the game of life. Its goal is growth and evolution because it is at one with the Big Dreamer.

Perhaps you, too, have made or were about to make a decision based on some ego demand or urge, the part of us that becomes afraid, that lusts, that rationalizes, and worries what others will think. Something inside said you would live to regret the decision if you followed through. At that moment you were in touch, however briefly, with your subconscious mind.

One mind exists, and it is divided into a number of gradations: The mind of the Big Dreamer, which encompasses the entire medium that we call mind, the collective unconscious we share with all humanity. Our individual subconscious minds or souls. And finally, our conscious minds, also known as the ego mind, which has an unconscious part which contains the memories and the programming of this particular lifetime.

The conscious part of our ego mind is our objective mind that by definition has self awareness. It is the part that tricks us into thinking we are separate. But we are not separate. From our individual perspective, life may appear to be chaotic and random, but everything is coordinated at the subconscious level. We each have our roles. Things click along when we are going with the flow. However, when we get off track by refusing the call, things go awry, and life gets messy.

Perhaps you know someone, as I do, who married a person he knew deep down he was going to divorce. Unfortunately for the person I know, the feeling didn't poke through into conscious awareness until the day the invitations were mailed. Even so, there was still plenty of time to call off the wedding. But he didn't. Two years later, after he and his bride had split, he came to the realization he'd—his ego had—talked himself into going through with the marriage because he didn't have the courage to tell the girl or his friends and his parents and her parents that the marriage would be a mistake. Embarrassment is typical of an ego concern. Had he listened to his inner voice, he would not have lived through the nightmare that ensued. But his ego mind blocked communication through the mechanism of denial because the truth was not what it wanted to hear.

The late Elisabeth Kübler-Ross [1926-2004], the Swiss-born physician and author of the perennial best seller *On Death and Dying,* was in attendance and helped ease the deaths of scores of patients. She studied the near death experiences of many more. She spoke of her own mystical, out of body experience and is generally acknowledged as one of the world's leading authorities in this area. She came to the conclusion that this inner voice is very real. In a 1977 lecture given in San Diego and published the same summer in the *Co-Evolution Quarterly,* she said, "If you listen to your inner voice, your inner wisdom—which is far greater than anybody else's as far as you are concerned—you will not go wrong and you will know what to do with your life." It is too bad my friend had not been exposed to these words, or if he had, that he did not heed them.

How can we get in touch? By answering the call to be "born from above." By recognizing that we are a part of the whole and that we have a conduit to the mind of the whole within us. As this becomes real for us, we move into the Kingdom of God—the condition of knowing and sensing our connectedness on a gut level. Over time our ego or lower self will come together in harmony with our subconscious mind. This cannot help but happen, and when it does we will experience the major payoff of our struggle upward on the spiritual path: a life where the pieces fit, where we understand why we are here, our purpose and how we are to achieve it. Fear will fade and finally vanish. As time passes, we become patient, collected, and serene. We are able to live in the eternal now and, perhaps for the first time, truly to experience and enjoy life in the physical world.

We all have egos, the part of us that has developed in this life from an unfocused awareness in our early days in the crib to the part of us that contains the memories of this life. It is the part that worries, that fights for life, for achievement, for glory and for recognition. In contrast, the subconscious mind or soul is not concerned with the trappings of the physical world. It seems to have been around for a long, long time, since the epoch of mankind's evolution from a species driven by instinct into a species characterized by self-awareness and free will. It does not experience fear or worry because it knows it will continue to exist throughout eternity. It possesses no desire whatsoever for self-aggrandizement.

This fits nicely into the theory of morphogenetic fields advanced by Rupert Sheldrake. Life itself has a morphogenetic field that first became differentiated from the overall field when DNA formed into one-celled creatures. This field evolved and changed over the eons as life took ever more complicated forms. Along the way, different parts of fields followed different paths of evolution. What has become my field and what has evolved into your field followed the path of primates. Each of us has a corner of it—figuratively, since the field is everywhere at once like TV transmissions. Ours dates from when we became differentiated from other primates via self-awareness. Your field, which is also known as your soul, just keeps on evolving as it continues reincarnating time after time.

Besides fitting this theory, the evidence for reincarnation compiled by researchers such as Ian Stevenson and Jim B. Tucker of the University of Virginia is simply too compelling to ignore. Readers who are Stage Two Christians may be put off by this, or dismiss it out of hand because it's not part of the doctrine of the Church. Nevertheless, I believe that Jesus and his contemporaries—both Jews and pagans—took reincarnation for granted, just as Hindus and Buddhists do today. Read the Gospels with this in mind. You will see that passages that once seemed obtuse snap into focus before your eyes. Many scholars think that as the canon of the Church was formalized in the Fourth and Fifth Centuries the concept of reincarnation was judged to be counterproductive. It was thought that some potential converts would resist or delay accepting Christ because they'd think they'd have opportunities in future lives. Reincarnation was doctrinally eliminated as a result.

But consider a few examples why it seems likely that Jesus and others of his time believed in reincarnation. For example,

John the Baptist was supposed by many to be the prophet Elijah reincarnated. Jesus himself said this was so. (See Matthew 11:14.) Once, Jesus asked his followers who people thought he (Jesus) was. They replied that many believed him (Jesus) to be one of the prophets—presumably reincarnated, since the last prophet had lived about 400 years earlier. Also, consider the story of Jesus healing the blind man as recounted in John 9:1-12, which begins as follows:

> *As he went along, he saw a man blind from birth. His disciples asked him, 'Rabbi, who sinned, this man or his parents, that he was born blind?'*
>
> *'Neither this man nor his parents sinned,' said Jesus, 'but this happened so that the work of God might be displayed in his life.'*

Since the man was blind from birth, the only way his sins could have caused his blindness was for him to have sinned in a former life. Jesus did not tell his followers this wasn't possible. To the contrary, he seems to have assumed it was possible, although he gives another reason for the man's blindness.

In researching reincarnation, I've found that libraries are well stocked on the subject. Since becoming interested in this subject, I have met and come to know well two different people who make their livings by helping others remember past lives and then release buried memories that are holding them back. In some cases, thousands of years have passed since a debilitating incident took place. I've visited the School of Metaphysics in Missouri and watched trained readers of the Akashic records report on past lives of workshop attendees. Additionally, I've read four books written by different past life therapists and edited a fifth. Rather than relate what is contained in those, however, I will give you a quick summary of a case reported in the 1988 book, *Many Lives, Many Masters.* I've chosen this because the author, Brian L. Weiss, M.D., cannot be accused by anyone of being a Looney Tune. He is a Phi Beta Kappa, magna cum laude graduate of Columbia University who received his medical degree from Yale, interned at New York's Bellevue Medical Center, and went on to become chief resident of the department of psychiatry at the Yale University School of Medicine. At the time of the case covered in his book, he was head of the department of psychiatry at Mount Sinai Medical Center in Miami Beach.

Weiss is a medical doctor and a scientist who has published widely in professional journals. Ethnically Jewish, he was a skeptic who had no interest in reincarnation. He was fully aware that most of his colleagues in the field do not believe in such things and waited six years before giving in to the feeling that he had an obligation to share what he had learned. He had much more to lose than to gain by telling the story of the woman called Catherine (not her real name) who came to him in 1980 seeking help for her anxiety, panic attacks and phobias. Read his book. I'll hit only a few highlights.

For eighteen months, Weiss used conventional therapy, which means that he and Catherine talked about and analyzed her life and her relationships. When nothing worked, he tried hypnosis in an effort to find out what she might be repressing that would account for her neuroses. Forgotten events in her childhood, in fact, were revealed that seemed to be at the root of several of her problems. As is customary in this type of therapy, she was instructed to remember them after she had been brought out of the hypnotic state. Dr. Weiss discussed what had been uncovered in an effort to dispel her anxieties. But as days went by, her symptoms remained as severe as they had ever been.

He tried hypnotism again. This time he regressed her all the way back to the age of two, but she recalled nothing that shed new light on her problems. He gave her firm instructions, "Go back to the time from which your symptoms arise." Nothing had prepared him for what happened next. She slipped into a past life that took place almost 4,000 years ago. Weiss was astounded as she described in detail herself, her surroundings and others in that particular life, including specific episodes, and in later sessions entire lifetimes, which seemed to be the root causes of problems. In all, she said she had lived 86 times. This, by the way, indicates she was a very young soul in that most who have gone through this type of regression therapy have lived hundreds of times.

Weiss continued using hypnosis in an effort to rid Catherine of her neuroses. In weekly sessions that spanned several months, she recalled and recounted in detail the highlights of twelve previous lifetimes, including the moment of death in each. People who played a role in one lifetime often reappeared as someone else in another, including Dr. Weiss himself, who had been her teacher some 3500 years ago.

Catherine had not had a happy existence over the last forty centuries. The overwhelming number of memories from her past lives were unhappy and proved to be the roots of her present day symptoms. Bringing them into her consciousness and talking about them enabled her to recover. Considering the number and intensity of her neuroses, psychotherapy would normally have lasted years before she was cured. In fact, her symptoms disappeared within months. She became happier and more at peace than she had ever been.

Weiss is an experienced psychotherapist who has dealt with thousands of patients. He is convinced that Catherine was not faking. She was unsophisticated and of average intelligence, a young woman who made her living as a laboratory technician. He thinks it quite impossible that she could have pulled off such an elaborate hoax and kept it up every week for months. Think about it. She was a physically attractive twenty-eight year old woman of average intelligence. She had a high school diploma and some vocational training. Could she have faked her neuroses? Could she have faked gradual improvement from one visit to the next, all the way to a state of being completely free of them? It hardly seems likely. Also, and this is where the plot thickens, she conveyed information about Weiss's father and an infant son, both of whom had died. Weiss is convinced she could not have known anything about them through normal channels.

This message from the other side leads to what some may find the most amazing aspect of her story: the spaces between past lives. Once, after having been murdered, she floated out of her body and was reborn very quickly. At the end of her next life, she described an experience remarkably similar to that related by thousands who have been clinically dead and come back to life. She rose out of her body, felt at peace, and was aware of an energy-giving light. It was at this time in this session that spirit entities spoke through her to Dr. Weiss for the first time. In a loud, husky voice and without hesitation Catherine said, "Our task is to learn, to become God-like through knowledge. We know so little. You are here to be my teacher. I have so much to learn. By knowledge we approach God, and then we can rest. Then we come back and help others."

Although Catherine was able to recall her past lives after she was brought out of a hypnotic state, she was never able to recall, nor was she particularly interested in remembering, the conversations Dr. Weiss had through her with several different spirit entities. These "masters," as he came to call them, spoke through her primarily for his benefit and only indirectly for hers. I will not go into detail about these exchanges; you may wish to read this book for them. Essentially, they told him that we incarnate into the physical world to learn what cannot be learned on the nonphysical plane. In that realm, whatever is felt or imagined instantly appears real or greatly magnified. The slightest ill will toward someone becomes rage. The smallest feeling of affection turns to all encompassing love. If you imagine a demon, a thought form of it will suddenly materialize before you. If you picture in your mind a lovely sunset viewed from a secluded beach, you will find that you are there. It is because of this that we need the thickness of matter. Matter slows things down so we can work them out. Earth is a school. The most important things we come here to learn are charity, hope, faith and love, as well as to trust and not to have fear.

Funny. This sounds an awful lot like Jesus or the Apostle Paul, doesn't it?

Let's leave Dr. Weiss for the moment and dig into the workings of reincarnation. What you are about to read at first seemed as fantastic to me as it may to you. Nevertheless, like Dr. Weiss, I feel compelled to share .

When the Life Force or spirit is withdrawn from anything, be it an animal, plant, person or object, the Life Force continues to exist, but the object it supported no longer is animated by it. That thing ceases to be alive—informed by the Life Force—and turns to dust. This is true of what we normally consider living things such as plants and animals, and it is true of what we may have thought until now were inanimate objects such as rocks, moons, mountains. Although the process of decay and return to dust takes longer for the latter, it will nonetheless happen in time when the Life Force is no longer present.

As Claire DuMond came to know in my novel, *The Secret of Life: An Adventure Out of Body, Into Mind,* the secret of life is the "urge to become," which in this book I've called the Life Force or subjective ground of being mind that pushes toward growth and is the opposite of entropy. In other words, life is imbued with an organizing ability that pushes it to evolve into ever-higher forms. Finally, when the sensation of being separate comes about, a soul—or an individual subconscious mind—is born.

Your subconscious mind or soul may have evolved on Earth, or it may have evolved elsewhere. As for myself, I've had a flash of memory that might be compared to an epiphany, which I believe is an indication of my own personal history of evolution. This moment of recall lasted perhaps for 30 seconds during which I "relived" all my pre-human lives in rapid succession, from life as a fish-like creature in some primeval sea, through reptilian forms, to a furry lemur-like creature that lived in a tree. This seems to fit. When a soul has learned all it can in one form, it seeks a new experience that will allow it to continue its upward push. Ultimately, it will grow and develop until it reaches a perfected state.

We live in a multi-dimensional reality, though under normal circumstances our physical senses allow us to experience only height, width, depth and the passage of time. Souls are evolving in other dimensions, and they are evolving on other planets in other solar systems of this universe. While recognizing that your soul could be much older than life on earth, one view of how souls evolved along with life on this planet will be presented in the paragraphs that follow. This is by no means intended as the last word on the subject and is meant only to give you one plausible explanation of the evolutionary path a soul may have followed.

A theory accepted by some followers of eastern religions is that souls that began their journeys here have been around in some form since the beginning of life on Earth.

They did not become differentiated, however, until the epoch recounted in the myth of Adam and Eve. Scientists would probably estimate this to have occurred about 200,000 years ago. As mentioned earlier, this was when we reached a point in the evolution of our minds that we were self-aware. We saw ourselves as different, separate and distinct from the rest of nature. Unlike birds and animals of the forest or the savanna, we no longer relied on nature and instincts to direct our behavior. Our minds could override what instinct said to do.

This is what the story of Adam and Eve is about, the development of objective awareness and the splitting off or separation from the field that resulted in self-awareness and free will. God told Adam and Eve not to eat the fruit of the tree of knowledge of good and evil. The snake, which represents Adam and Eve's all too human nature, or ego mind, said to go ahead and eat. Rather than consult God before taking action, Adam and Eve acted as humans usually do today, and proceeded to do as they pleased. By exercising free will in this manner, they severed their connection to God, and humankind has been suffering the consequences since. We have, in effect, cast ourselves out of the Garden, with the result that we are no longer able to tap effortlessly into the abundance that nature is always ready to bestow on us.

The way back, of course, is to reestablish a relationship with our personal subconscious mind, or soul, and thus our connection to the transcendent. But this is a digression. The point here is that the origin and evolution of souls in this particular scenario would have followed the course of evolution from one-celled animals in the sea, to creatures who first walked on land, to tiny mammals, to pre-apes, to homo sapiens. It was as homo sapiens that we became differentiated. The Adam and Eve step was absolutely necessary. But we've been on a plateau now for about a hundred thousand years. Since the course of evolution is a spiral rather than a straight line, our next step is to return to the state the first man and woman enjoyed, but on a higher level. The time has come for many on earth to reconnect with the whole, while remaining aware of their separateness and maintaining free will. Metaphorically, they will return to the Garden and this time remember to keep God in the loop. When this happens, the bad times will be behind us. Our every desire will be fulfilled.

Some will argue that reincarnation isn't likely because there are so many more people alive today than in the past. The population of earth has exploded in recent centuries. If humans aren't humans unless they have a soul, and if a human soul must be built up through many incarnations, then where, they wonder, did all these souls come from?

It seems to me that there are at least two possible explanations. One is that souls which evolved on earth are incarnating more frequently now than in the past. In other words, they are spending less time in the state between lives. One therapist who has helped many recover from psychological problems due to past life traumas indicates that with respect to his clientele, the duration between lives ranges from a high of 800 years to a low of ten months.

Another possibility is that souls are pouring into this world from all over the place, that evolution of these souls up to this point may have occurred elsewhere for many alive today. Morphogenetic fields are composed of information as opposed to energy. According to quantum physics, unlike energy that must travel and diminishes in intensity with distance, information is everywhere in the field at once. It is nonlocal. (This, by the way, coincides with Thomas Troward's theory concerning thought and life or spirit covered in a previous chapter.) Since a thought form is everywhere at once—nonlocal—souls can come from anywhere in the universe; no travel time is required.

Parenthetically, we can expect life forms that have evolved on other planets to be similar to those on earth, provided physical conditions of the planet are similar. I say this because this is true of life forms that evolved in Australia after that continent became separated from the rest of the world's land mass. We see equivalents of dogs (dingoes) and cats and other animals in the land down under that are not exactly dogs or cats. They are marsupials, not mammals. So on a family tree, they would be placed closer to the opossum than to the animals they resemble. Water prevented the spread of genes of cats and dogs but not their morphogenetic fields. These are everywhere at once—unhindered by water. These fields have influenced the shapes and forms of Australian creatures—though their genes may be quite different. When you think about it, those strange creatures might just as easily have evolved on another planet since physical contact with the rest of the world had been cut for millions of years.

Let us now consider the process of evolution of a human soul. When a new human soul starts out, the number of foolish or evil actions, thoughts and words the soul is responsible for far exceeds those of the good variety. This is understandable. It is also where the law of karma comes into play, as a basic learning tool provided by the universe. According to the law of karma, which boils down the law of cause and effect, every thought, word or deed must produce a definite result, good or bad, and the result must be felt by the person responsible. Experiencing the law of karma is one of the ways we learn. It is one reason many lives in physical form are necessary. We usually do not live long enough for all our acts and deeds to play out in a single life.

The Bible tells us, "As you sow, so shall you reap," or as an old friend of mine in the ad business was often heard to say, "What goes around comes around." I believe that Jesus was talking about karma, for instance, when he said:

> *Do not judge, and you will not be judged. Do not condemn, and you will not be condemned. Forgive, and you will be forgiven. Give, and it will be given to you. A good measure, pressed down, shaken together and running over, will be poured into your lap. For with the measure you use, it will be measured to you.*
> —Luke 6:37-38

A selfish act on your part that causes misery to someone else earns a unit of bad karma. This must be repaid by your suffering from a similar action at the hands of another person, either in this lifetime or in one to come. A kind act on your part earns a unit of good karma. The result of this action can be either the erasing of a unit of bad karma or experiencing the same amount of kindness from someone else. You might say that karma is the metaphysical law that's equivalent to Newton's law of physics—for every action there is an equal and opposite reaction.

When I first learned about this and the realization of it took hold, I began to think back over my life and to remember things I'd done to others that had caused them pain. A number of instances of thoughtlessness, and two or three of outright cruelty, came to mind. I truly felt remorse and was thrown into a kind of depression. It was as though a black cloud hung over me and my future. I could almost feel the pain I'd caused and went about wondering how I'd ever repay these debts. At that time, I didn't know the therapeutic and practical value of confessing my sins directly to God, or to Christ, and asking forgiveness. Rather, I was convinced that I was doomed to suffer the same level of misery that I'd inflicted.

Let me interject here that this happened thirty years ago. I have since come to realize that karma is not a form of justice or punishment. It is a teaching tool. Once a person has learned the necessary lesson, the associated karma is dissolved. This brings up the subject of a personal God verses the creative force that I have called the non-dual, subjective ground of being mind. Edgar Cayce's readings suggest that both exist. It seems to me our personal God may actually be our own evolving soul or Higher Self.

Let's get back to my dilemma about repaying karmic debt. One day, I was running along the Canal of Burgundy during a summer sojourn to France. I came to a stop and said, "Please, God, Please. Even out the score. Give me a level playing field. Make whatever needs to take place happen so that these debts are paid."

A Jewish friend later told me that Jewish prayers to wipe the slate clean include the phrase, "But not through pain or suffering." I thought, "Now he tells me." I learned firsthand that you get exactly what you ask God for when what you ask will result in spiritual growth. In this case, what I asked for started coming three days later when I was back in the States. I took my daughter's brand new ten speed bike for a test spin down the hill in front of our house. My foot slipped, caught on the pavement, and the metal pedal completely severed my Achilles tendon. The result was a ghastly wound. I spent almost two weeks in the hospital, had two operations, suffered a great deal of physical pain, and was in a cast from the tip of my toe to the top of my thigh for eight weeks. It took nine months before the wound was completely closed, then another nine before I was able to walk without a limp.

But that was not all. While I was still in the cast, my wife announced that the spark had gone out of our marriage. She was sick of living in the United States. She was leaving me, filing for divorce, and taking my daughter with her to live in France.

As a parent, I can imagine that the death of a child may possibly be the worst possible experience one can endure. If this is so, then having your only child, age twelve, taken to live 3,500 miles and six time zones away is number two. It was not a good year, but at least the slate was wiped clean. As you might expect, I developed spiritually. Adversity is a great teacher. The experience also taught me that it is possible to get in touch with the Big Dreamer and to have a request granted. I do not advise you, however, to follow the same course because I now believe the process of arriving on a level playing field with respect to karma does not have to be so painful. This is partly due to the fact that the purpose of karma, the law of cause and effect, is not retribution in and of itself. The Big Dreamer finds no joy in extracting "an eye for and eye and a tooth for a tooth." Rather, as with most things connected with the subconscious mind and universal law, the purpose of karma is to foster spiritual growth. Sometimes an "eye for an eye" is the only way to make a point. This is especially true during the early part of our spiritual journey. Then, the only way we will fully understand the consequences of our actions, our thoughts, and our intentions will be to experience those consequences firsthand. When O. J. Simpson comes back in another life as a woman and is brutally murdered by a bigger, stronger person with a knife, chances are he will finally "get it" deep down in his soul.

The end goal of the law of karma, you see, is the shift in consciousness that I've been harping about. It is the "aha!" experience, the gut-level realization of what one has done when one realizes that he and others are the same person—not literally at this moment, perhaps, but certainly at another moment, or in a different incarnation. This is likely what Jesus was driving at when he took a child in his arms and said, "Whoever welcomes one of these little children in my name welcomes me; and whoever welcomes me does not welcome me but the one who sent me." (Mark 9:37) And, when speaking of the hungry, thirsty, and the downtrodden, he said, "I tell you the truth, whatever you did for one of the least of these brothers of mine, you did for me." (Matthew 25:40)

In these quotations, Jesus is saying we are all from God and of God, and we all have played or will play every role and part in the dance of creation. This being the case, to help or harm another is to help or harm yourself at some point in your past or future. It is also to help or harm God.

How is possible to help or harm God? There is only one screen of awareness on which the moving picture of reality plays. In aggregate, the total screen is God's awareness. Each person's awareness, on the other hand, is a tiny sliver of the larger screen.

Does one who has committed murder in this or a previous life have to experience the same violation? I believe that meditation, study, reflection can lead to a higher level of consciousness that, combined with true repentance, can avert having to live through "eye for an eye" retribution. The universal subjective mind doesn't judge and it doesn't hold a grudge. Being subjective, it doesn't even sit around and analyze. If retribution is no longer needed for growth, it will not happen. This is one of the key messages of Jesus. He came to show the way, his way, to Christ-consciousness and everlasting life. The attainment of the Christ-consciousness triggers the "law of grace," and this dissolves the need for a karmic boomerang. Spiritual consciousness "fulfills the law," to use Jesus' phrase, in the sense that it annuls the erroneous thinking that was the source of the wrongful action. "I come not to destroy the law," is what Jesus might have said "but to teach you how to fulfill it through an elevated spiritual consciousness."

Full attainment of this consciousness is not easy. "Remember," Edgar Cayce said in one of his readings, "there is no shortcut to a consciousness of the God-force. It is a part of your own consciousness but it cannot be realized by a simple desire to do so. Too often there is a tendency to want it and expect it without applying spiritual truth through the medium of mental processes. This is the only way to reach the gate. There are no shortcuts in metaphysics, no matter what is said by those who see visions, interpret numbers, or read the stars. They may find urges, but these do not rule the will. Life is learned within the self. You don't profess it; you learn it."

How, specifically, does one learn it? Regular meditation and prayer. Fasting. Study of the Scriptures and books such as this. Rendering service to one's fellow men. All these will be helpful. Depending on where a persons stands in the evolutionary process, regular practice of these acts may be all that is required. But I personally have been at it forty years in a concentrated way and can say with honest conviction that I still have a good way to go. Every now and then when things seem to be going wrong, I suffer bouts of fear and doubt. Intellectually, I know that this is counterproductive. If I were fully evolved, I wouldn't have those doubts. Even so, I am putting my talents to use, and I do feel fulfilled in this regard. I have a solid, comfortable, affectionate marriage. I share and revel daily in the joy of healthy, happy, well-adjusted children. My greatest shortcoming seems to be the almost subconscious fear that if I don't push myself, the good things the universe has flowing in my direction will somehow stop coming. Perhaps this stems from growing up poor. Nevertheless, I'm making progress. I have faults, but I know it. Some would say I'm patient, but I know I'm too quick to anger. I'm making an effort to correct this. If I stay at it, perhaps full mastery of life may not be all that far off. This life? The next? The one after that?

I believe I may have entered this life at a level of evolution higher than I now have achieved. But for the first half of my days in this incarnation, I was sliding backward, losing ground. Having come to my senses, I've been able to make up much of the lost ground quickly because in effect, all I've had to do is "remember" (re—member) what I already knew and recapture the level of consciousness.

Backsliding is a potential danger an old soul must face when he or she incarnates. But as is always the case, the consequences of the ill-advised exercise of free will be used by the Big Dreamer as an opportunity for spiritual growth. The experience of having to crawl back, for example, helped prepare me to write this book. If I had come into flesh with prior knowledge intact, I'd have faced two problems in sharing it. First, it would have been a priori, or part of me and my makeup, and therefore difficult to verbalize. Second, It would not have been gained in the white-hot crucible of a society of doubters and skeptics. But because it was acquired gradually, over the years, I can present it with a good idea of what objections will need to be overcome. After all, I had to overcome them myself before I accepted the knowledge as truth. Perhaps this makes what I have to say more convincing.

Having regained what I have to share in this lifetime, I can do so in the context of this lifetime and the time in history that we share. My sincere desire and intent is to help you, too, to re—member what you once knew, and to help you soar to a new level of understanding. But let me quickly add that growth cannot be mechanically induced. Unless and until the heart is sufficiently tenderized, practicing charity, for example, will be in the Apostle Paul's apt phrase, "as tinkling brass." The rich man who gives it away with the idea of buying his way into heaven has bought nothing. It's what is in the man's heart that counts. What he truly gives, he keeps, because of the joy he feels. But don't misunderstand. This doesn't mean he should not give, even though he does not feel joy. Perhaps his deed may be the act his soul needs to start it on the proper path. But this won't let him skip a grade. Souls in kindergarten, spiritually speaking, cannot jump ahead to college.

Let's think for a moment about the new soul with very little understanding. In the early incarnations, this entity will go about creating much havoc. Fortunately for him or her, no one is expected to suffer more in one lifetime than he can stand. Units of bad karma that aren't worked off by good deeds or a poke in the mouth are carried forward to be worked out in future lives.

Think about a person you know or have come in contact with who has what you now recognize to be a young soul. The individual appears to have very little conscience. He or she simply cannot hear the still small voice and thinks nothing of spraying his housing project with bullets just for the fun of watching broken glass tinkle to the ground. During the early incarnations, a person like this will pile up more karmic debt than he works off. But as time goes by and he continues to incarnate, the connections between his subconscious mind and his ego grow. Once the lines of communication are open, conditions improve. At last, the right wavelength is found and the mother ship comes in loud and clear. This is why a person with an old soul appears to have a highly developed sense of right and wrong. When this happens, bad karma stored up from early lives is whittled down. Many people today have reached this state. You probably are one. If so, a short jump is all that's required to make that shift in consciousness.

This book isn't meant for those who possess young souls. Past-life therapists tell me that often such individuals incarnate with no life plan. Their lives are chaotic at best and purposeless at worst. They aren't likely to tap into the universe's cornucopia, nor does it seem likely they would have much interest in what this book has to say. On the other hand, having stuck with me this far, I suspect you have a relatively old soul and may be on the cusp of a breakthrough. This being the case, it's highly likely that you came to earth with special talents to use to the benefit of humankind and have an important purpose or goal to accomplish. If you have achieved Christ-consciousness in a previous life, your visit to earth this time is to help others do the same. If you have a way to go, it is to face and overcome certain problems or to work out karma from a previous incarnation. Soon we'll cover how to drop the karmic baggage that's holding you back so that you can enter the flow.

Speaking of flow, I've heard a number of people who profess to be New Consciousness thinkers repeatedly make statements to the effect that "everything is working just as it should" at any given moment. Perhaps this is to remind themselves not to be concerned with the outcome, but rather to concentrate their energy on doing what they are here on Earth to do. I agree with this approach. I've never made money, for example, by trying to make money.

And believe me, there are times when I've tried very hard. To the contrary, I've made a lot of money by concentrating my effort on what I do well and "just doing it," to borrow the Nike slogan. Each time I've let the chips fall wherever, things have indeed worked out to my benefit, provided I was doing what I now see I'm here to do.

Even so, I do not agree that everything is always working out for the best at any given moment. Though I rarely say anything, these kinds of statements sometimes rankle me because they are contrary to the law of free will. On the material plane, in other words here on Earth, each human being is free to choose, and therefore, is free to make mistakes. Dumb mistakes. Mistakes in calculations. Mistakes in judgment. We are free to be just plain cruel or stupid. Thomas Troward pointed out that the universe operates by laws. Violate a law, whether or not you even know the law exists, and you will suffer the consequences. If you don't believe this, watch what happens when an unsuspecting child sticks his finger in an electric socket. Metaphysical laws are as consistent as those of physics. For example, it is a law that what you believe will happen will eventually happen. So, believe you are going to go bankrupt, and you will.

Wait a minute, you may say. I've thought terrible things might happen a number of times, and they didn't.

The key here is that you "thought they might happen." You no doubt hoped they wouldn't. Though weak, hope is a kind of belief. It is a belief in possibilities. You believed this might happen or that might happen. You were sending the subjective mind mixed messages, and fortunately for you, hope was stronger and won the day. You'd have been better off to believe you would succeed and not take chances.

Some people think there's a fairy godmother looking after them, or maybe a guardian angel, and that's okay. It may be true. In fact, I think there are souls or guides beyond the veil who are with us on the journey of life. In a way, these disembodied beings are part of us and always with us. But I tend to think things usually happen because the law of belief is at work. It's a matter of maintaining consistency so that we are able to see and learn how life works. However, I also believe "divine intervention" happens as well, albeit rarely, because I've experienced what to my mind could have been nothing less. In an upcoming chapter, I'll tell the story of the divine intervention I experienced, but let me say here, I think it happens only when the consequences of it not happening would thwart Spirit's push toward growth and evolution.

It seems to me belief can bring about what appears to be divine intervention. For example, Jesus is said to have turned water into wine, to have walked on water, and to have fed 5,000 people with a few fish and some loaves of bread. Jesus didn't just believe, he knew he was in touch with and an extension of his Father in heaven. I've heard firsthand reports from people I trust—one the founding partner of a large law firm—of miracles similar to this being performed today. Turning water into wine is one.

It's important to realize, however, that things aren't automatically going to work out right. Edgar Cayce indicated that things can go awry even though they have been planned with the clarity that surely must exist on the spiritual plane. He assured us, for example, that the process of birth and rebirth does not always work as intended, and that sometimes errors are made. Cayce said that we choose our parents and our circumstances from what is available at any given moment. The circumstances may be far from perfect, yet we may elect to go ahead anyway and, figuratively speaking, keep our fingers crossed. The fallout may be that a soul may discover, after having made a choice and been born, that the parents are not living up to expectations they seemed to have offered before birth. Realizing that its own inner purpose for the incarnation may be frustrated by the altered circumstance, the soul may decide to withdraw. Cayce said that this may be the cause of at least some infant mortalities.

For detailed descriptions of what happens between incarnations, and the process of selecting parents and circumstances, read the books, *Journey of Souls,* by Michael Newton, Ph.D., Llewellyn, 1994, and *Life Between Life* by Joel L. Whitton, M.D., Ph.D., and Joe Fisher. The authors of both are past-life therapists. Both provide interesting information, although I found the one by Dr. Newton to be slightly more enlightening. Essentially, groups of souls numbering a dozen or so who are at the same level of development work in what might be called clutches. Each group has guides or teachers to help them. The guides are a level or two farther along the evolutionary path than those in the group and do not incarnate as often. These guides are at work virtually all the time, behind the scenes during an incarnation and at the head of the class in between—not unlike your high school math teacher. Members of the clutch often incarnate at the same time, in the same place, often in the same family, with the goal of helping one another accomplish specific tasks, work out karma, or learn specific lessons.

Soon after bodily death, a soul's guides or teachers, or what also might be described as elders—up to three, according to Dr. Whitton in Life Between Life—will expose the newly returned entity to a detailed review of his or her just-completed life. These elders do not judge the entity, however. The entity judges him or herself—actually feeling the pain or the distress, as well as the joy, that he or she may have caused others. The elders or guides make comments and suggestions. They are non-judgmental in their approach and often provide comfort to the entity, who may find the review gut-wrenching. This is very likely to be the case if a number of mistakes were made, or if opportunities were wasted that might have led to the accomplishment of goals that had been set forth for the entity's life.

This seems a good spot for a couple of anecdotes that relate to this. When he learned I was writing this book, a friend told me something that happened to change his life when he was a teenager. He had an unhappy home and was deeply depressed. His mother was wacky, and she and his father constantly argued. He had almost no friends. It was so bad, and things appeared so hopeless, that he was seriously contemplating suicide. One night, after he'd gone to sleep with suicide on his mind, he had the sensation of being shaken awake. He opened his eyes and saw two strangely clothed men. They grabbed him and whisked him upward, directly through the ceiling and the roof—or so it seemed. He now realizes he had passed through the tunnel we've heard so much about into a kind of limbo where the three had a chat.

"Don't do it," one of them said.

"Do what?" he asked.

"Don't commit suicide."

He looked at the man sheepishly and frowned.

"It won't do you any good," the other said. "If you do, you'll just be sent back again, and again, until you get it right."

My friend understood precisely what they meant. If he ended his life prematurely, he'd have to be born into the same circumstances over and over—like the character in the movie *Groundhog Day,* who had to live the same day again and again until he finally got it right.

Another friend, a therapist with a Ph.D., uses past life recall to help his patients get over their phobias and neuroses. He has explored his own past lives, and says he's been able to remember back about 15,000 years to the time before he came to this planet. At that time, he was engaged in an intergalactic war—sort of like Star Wars, I guess. He and his crew were captured just as they were about to blow up a planet inhabited by the enemy. He and the crew subsequently were sent to earth for incarceration and rehabilitation, and he's been here ever since, forced to reincarnate over and over again. He'd like very much to break out of the cycle but, so far, hasn't been allowed. He believes he is part of a small percentage of the population of our world who are prisoners in a penal colony. For them, Earth might be compared to Devil's Island in the South Atlantic off the coast of French Guiana, which was a French penal colony from 1895 to 1938. Instead of criminals from France, however, prisoners like him from other worlds are put here to "get them off the streets," so to speak. He's calculated the total number of prisoners on earth at about five million.

Returning to our description of the inter-life, quite a bit of time may be spent between incarnations, during which an entity will study his most recent past life as well as other lives he or she has lived. As souls become more highly evolved, they tend to spend more time between incarnations than do less evolved souls. They are more cautious about picking the circumstances of upcoming lives. Also, compatible parents don't come along as frequently for the more evolved souls. In the meantime, the soul may use the time to hone particular talents or skills.

Once back in flesh on Earth, a soul may accomplish what was planned for a particular incarnation more rapidly than anticipated. This was the case for one of Dr. Whitton's subjects. With nothing left to accomplish in the lifetime, some sort of premature death would normally have occurred so that the soul could return for rest and recycling. Instead, the living soul was given a new assignment and was allowed to stay on Earth and continue evolving. This way, the remainder of the incarnation was made fruitful and productive, and what would have been viewed as an untimely death by us mortals was averted.

It is important to know that taking on a new assignment is possible. This means we don't have to die just because we've accomplished the objectives set forth before birth. In effect, we can begin a new incarnation without having to go through the bother, drama and trauma of death, rebirth, childhood, adolescence, and so forth. If science indeed finds keys that unlock the mechanism of aging, this could become the norm, provided people understand that the purpose of life is spiritual evolution—

their own and others'. I say this because it seems only logical that people who decide to retire and play golf all day, everyday, will have a serious illness, accident or something else happen in order to get them back on track and productive once again.

Let's go back to the life between lives. When the time comes for rebirth, the panel of judges will review or help identify objectives and lessons for the next incarnation, perhaps giving the entity a choice of a couple of different families in which to incarnate. The soul must agree to the selection, although it appears that this agreement is often given reluctantly. The upcoming incarnation is then planned much as a writer might outline the plot of a novel. The elements must be in place and the supporting characters ready and waiting. The entire process can be tricky. Race, nation, region and familial circumstances are factors. Someone who was a bigot in a past lifetime may come back clothed in the race they were prejudiced against in order to experience the other side of the equation and to work out the karma they created.

Amnesia of the time between lives is important. If we know what's coming, the purpose will be defeated. Fortitude and courage, for example, will not be acquired if the harrowing experience and outcome are known in advance. For lessons to be internalized, they must occur spontaneously, without foreknowledge. So as a new ego begins forming, amnesia sets in—starting when a baby begins to bring its surroundings into focus.

Most of us who have been around babies and small children instinctively know that they have just arrived from some heavenly place. I could see this clearly in my son, Hans, when he was still a toddler. As long as his needs were met, including food and a lot of good, healthy interaction with people who love him, he appeared absolutely delighted to be here. His face almost seemed to glow. The quality I have in mind has been recognized by poets down through the centuries. It was best captured, in my opinion, by William Wordsworth in the fifth stanza of his poem "Ode." This work is perhaps best known by the subtitle, which is, "Intimations of Immortality from Recollections from Early Childhood."

> *Our birth is but a sleep and a forgetting:*
> *The Soul that rises with us, our life's Star,*
> *Hath had elsewhere its setting,*
> *And cometh from afar:*
> *Not in entire forgetfulness,*
> *And not in utter nakedness,*
> *But trailing clouds of glory do we come*
> *From God, who is our home:*
> *Heaven lies about us in our infancy!*
> *Shades of the prison-house begin to close*

Upon the growing boy,
But he beholds the light, and whence it flows,
He sees it in his joy;
The Youth, who daily farther from the east
Must travel, still is Nature's Priest,
And by the vision splendid
Is on his way attended;
At length the Man perceives it die away,
And fade into the light of common day.

Whether a little person "trailing clouds of glory" makes progress in this incarnation or whether he fulfills his destiny or slides backward, will depend in large measure on the efforts and abilities of his parents. Consider for a moment how important it is that we create a loving environment in which a child can flourish and develop and pursue "natural" talents. Indeed, we can make it easy or difficult and, in the process, create a good deal of negative or positive karma for ourselves. In a very real way, our parents are our guides at least until we're grown. It is an awesome responsibility.

All that our invisible guides can do is provide us with guidance when it is sought and help create favorable conditions for us to "pursue our bliss," as Joseph Campbell called it many times. Of course, the subconscious mind will remain a subpart of the universal subconscious doing its best to keep the communication lines open. But each of us is born with free will. We always have the option of going against what our intuition or "better judgment" tells us.

A major point of this chapter is that you are here for a reason. Your birth was not an accident. You have a choice. You can try to find that reason and live it, or you can do as you please, and perhaps get so far off track you'll never get back. You are a soul with a body, not a body with a soul. Maybe you have made some mistakes. If so, it may not be too late to correct them, especially now that you know you are the driver and not the vehicle.

It's hard to doubt that we are all here to evolve. Some do so rather quickly. For others, it takes longer. Some may never make it.

But why? What is the purpose behind it all?

In his book, *The Seat of the Soul,* Gary Zukav argues that the purpose is the rounding out and eventual perfection of the soul. He wrote, "When the soul returns to its home, what has been accumulated in that lifetime is assessed with the loving assistance of its teachers and guides. The new lessons that have emerged to be learned, the new karmic obligations that must be paid, are seen. The experiences of the incarnation just completed are reviewed in the fullness of understanding. Its mysteries are mysteries no more. Their causes, their reasons, and their contributions to the evolution of the soul, and to the evolution of the souls with whom the soul shared its life, are revealed. What has been balanced, what has been learned, brings the soul ever closer to its healing, to its integration and wholeness."

Let us ask the question one more time. Many believe it is to become co-creators. But is that the end of the line?

If we think about this long and hard it will occur to us that perhaps the transcendent, the universal subconscious "one life," has set about the task of reproducing itself. Think about it. As Richard Dawkins observed in his study of cheetahs and gazelles, propagation is an underlying theme of nature. Every organism from the smallest amoeba to the biggest whale has this as a primary objective. Why shouldn't the universe, the largest organism of them all? Perhaps at some point in our development, long after this life is finished, we will not only be the field, we will be a new reality.

Perhaps a less grand view is that we will be fully evolved beings whose function is to help in the construction of new realities, new universes. This idea was held by the occultist, W. E. Butler. He said, "We're going to be universe builders in company with God. We are going to be tools, instruments in the hands of the Eternal as His will prevails in the universes which He has formed and in which He lives, moves and has His being and which He is bringing back to perfection from their fallen state. And you and I have the privilege of being coworkers with Him and with the whole of creation which is part of His work."

Parting Thoughts

Going forward, as you take the steps you have set out for yourself and row your boat in the direction of the flow, I suggest you set aside a full day once each month first to develop, and then to update, add to, and refine your plan.

Remember, life is the dream of the Source, and you are a character in the dream. You have a role or roles to play. Before you arrived here on earth, you took them on, and you made a solemn vow to carry them out. By taking the time I've just suggested, you will hopefully recall that solemn oath and the vision you had before birth of the self-actualized you. Now you have a choice to make. You can either make good on your promise, or you can welsh. If you welsh, you will view the consequences when the time comes for your life to be replayed before you, your judges and your guides. So, go to a library, or some other quiet place. Take time to think, and plan.

Getting it right might not be easy. It will take courage. To get started, you must clear the junk from your mental attic. That's why you must set aside time to work on this with no interruptions. You must also take time to forgive yourself and others. You must get past your fears and re-place them with positive belief. You must learn to trust. You must commit to change. You must be willing to suffer hardships. You must give up the "certainties" of the world you have created up to this point in your life.

Once you have a plan, you need to stick to it. Devoting a half-hour once, perhaps twice each day to meditation and a full day each month focus on and update your plan. In between, keep listening to the still small voice. When you review your life, you will think about the decisions that brought you here. Were they the right decisions? Did you feel good after you made them?

Remember to try to discover why your soul chose the circumstances of your birth. You will think back to what you loved to do as a child. You will ask for guidance, and you will receive it. You will follow your bliss.

Begin to build trust in your intuition by following the direction you receive when making small decisions. Eventually you will follow the gentle voice you come to know inside as you make the big ones. This may be frightening at first. It may be frightening because you will not know if it is indeed leading you toward your desired destination. But after a while, after you have learned how to trust, not knowing will become part of the fun, like opening packages at Christmas. You will be on an adventure as thrilling as any attempted by Indiana Jones. You will be the director of your own lucid dream.

Or you may finish this book, put it down and forget about it. No doubt that is what many will do. They've spent their lives doing what others told them. They've carved out a place for themselves. It isn't all that exciting or fulfilling, and it can be difficult. But they've become comfortable with who and what they are. Why change? There's no proof they're in danger of getting so far away from their soul and their source that they may never find the way back. No conclusive proof can be produced that a nonphysical realm exists. No scientific experiment shows that any part of a human being survives death. Those who died and were resuscitated? Many scientists still maintain it was all in their heads. A trick of the brain; a lack of oxygen. Besides, it's so much trouble to change. And what would people think? Things are comfortable the way they are. Life isn't so bad. Why rock the boat? And this Martin fellow says that once a person gets started, he or she won't want to stop, and may end up changing completely, as though all they wanted was to remodel the kitchen and ended up rebuilding the entire house.

Sun room done? Now how about the den? Oh, and you need a wing off to the side. And a second-level master bedroom with skylights and a fireplace.

All the while this renovation is going on, dust and debris are piling up, and the person occupying the house has to live in the middle of it.

"Wait a minute, I had a nice little bungalow," you may be tempted to say. "All I wanted was a new kitchen. You're turning this into a mansion. When will the job be completed?"

The architect, your higher self, will shake her head and say, "Not for a very long time, I'm afraid. You'd best get used to it."

Maybe you don't want a mansion. If this is the way you feel, I doubt there's anything I can say now to change your mind. You might as well stick with the bungalow.

I'll tell you something from personal experience, though. There is no greater joy in life than doing what you are here to do. Getting there may be difficult, it's true. But if you listen and persevere, if you earnestly follow the path laid down, you will receive help. After a while you will begin to sense unseen hands guiding you and the way will become less difficult to find. The trials won't be as hard to bear. There will be blind alleys, of course. There will be disappointments. There will be tough lessons to learn, but gradually you will come to a gut level understanding of what your existence as a human being is about. You will come to a gut level understanding of how you fit into the scheme. You will feel at one with it all and yet maintain your sense of self. You will come to know what you are doing. You will see outcomes materializing well before they arrive. You will choose what to pursue and what to avoid.

When you arrive at this point, you will realize that you have come to power, spiritual power, and with this realization will come joy. Can you imagine the buoyancy you'll feel? Whether it's mastery of a sport such as tennis, mastery of the card game of bridge, a musical instrument or a foreign language, the arrival at the state of really knowing what you are doing always brings joy.

And abundance. Not so much in the form of money per se, but the true riches of the universe, which will flow effortlessly to you because you are working with the universe, instead of rowing upstream.

And health. Your body will respond to the new life you've found. No longer will there be any reason for aches and pains. No longer will there be any thought of or reason to contemplate the possibility of death, for you are on the path to Eternal Life. You will be a vibrant, living cell in the larger body of humankind, fulfilling your purpose and your promise. You will grow every day and help others do the same.

Yet with all this will come a sense of deep humility because you will know that it is not you who brought you here. It is your subconscious mind and the mind of the Source. Perhaps there will be some small pride in knowing that you finally have learned how to listen. But you will be careful to guard against feeling a sense of pride in any form. One of the lessons you will have learned on the way is that support is withdrawn from those who believe they are accomplishing great things on their own. The saying, "Pride goes before a fall," is true. The prideful soon learn how little they can accomplish on their own.

There will also come a sense of aloneness. Not loneliness, because you will have friends, you will have family, you will have others who are with you on the path. But few, if any, will have arrived where you have arrived. Few will be the number with whom you can share your feelings and insights. Few will understand them completely. If you want a sense of what this is like, read the Gospels. Time and again you will witness the frustration Jesus experienced. Often, even his closest Apostles could not grasp the true meaning of his words—for Jesus knew that he was at one with the Source, whom he called his "Father," and everyone around him thought they were separate, isolated, and robotic-like—as the vast majority of people think they are today.

In spite of this, you will possess a new, deep understanding of your true worth. It will be impossible to continue to think of yourself as meaningless or insignificant once you understand and take to heart that you and the Source are one, that grace exists for you, that guides are constantly with you, and that you are in truth the very Force of Life itself seeking expression, self-awareness, and understanding.

In closing, let me say that my wish, my prayer, my hope is that you will put to work your unique talents and gifts for the benefit of the greater whole of creation. My hope is that you will answer the call to adventure when it comes, and in so doing enter with me into what Jesus called God's Kingdom. Believe me, you will not be sorry in the long run that you did.

I feel gladness for you and joy that I might have been a conduit in bringing a deeper understanding of yourself. Now you know who and what you are, that you came here to pursue a specific goal, and that possibilities are unlimited and boundless. My sincere wish is that my efforts have helped you move another mile on the journey. Thanks for taking that mile with me.

Keep moving ahead. Remember always to look for the light. Expect it to be there, and it will be. Go for it.

And remember always: As you believe, so will it be for you.

# # #

Stephen Hawley Martin
and Other Books He Has Written

Stephen is an author, ghostwriter and publisher. If you have written a book you want to have published, or if you need editing assistance or ghostwriting services, please get in touch with him and begin a conversation. You can do so through his website: www.shmartin.com

A few other books by Stephen you may enjoy can be found on the pages that follow including the first two books in this series.

Life After Death, Powerful Evidence You Will Never Die

Second Edition

Stephen Hawley Martin

Will your consciousness continue after you die? Read this book if you want to be certain it will. This highly-acclaimed book presents the findings of research that began at Duke University in the 1930s and continues today at The University of Virginia. More than a hundred experts were consulted such as quantum physicists and researchers into the true nature of reality.

Kindle: ASIN: B06W572NRB
Paperback: ISBN-10: 1543134327

Life After Death
Book Two,
Achieve Joy Now &
Bliss in the Afterlife

Stephen Hawley Martin

Once you have reviewed the evidence and have beomce conviced that you are an eternal spiritual being having a temporary physical experience you will want to read the this book. In it, the author descibes what it was like to merge with the Cosmic Mind, and he shares insights he brought back with him that you can use to find joy in this life and bliss in the next.

Kindle ASIN: B09GHY3SRN
Paperback ISBN: 979-8479020360

All the knowledge of the universe resides within you because at a deep level all minds, past and present, are connected. Everything that has ever happened, every thought, every idea is there. The trick is to draw out information when you need it. In this book Stephen explains how he learned to do so and how you can, too.

Kindle: ASIN: B07HHFFWP8
Paperback: ISBN-10: 1723835250

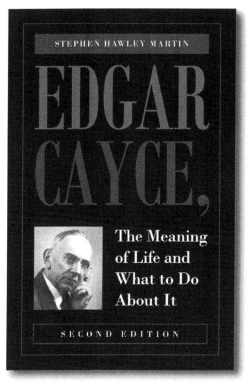

"Spirit is the life, mind is the builder, and the physical is the result." Repeated often by Edgar Cayce [1877-1945] when in trance, these words describe the formula that creates all that is, including your personal reality. Known in his time as "the Sleeping Prophet," the accuracy of the information he revealed that could be verified was absolutely extraordinary.

Kindle: ASIN: : B08W5DL6MD
PB: ISBN-13: 979-8706169121

THE THREE KEYS TO LASTING HAPPINESS

AND HOW TO OBTAIN THEM

Stephen Hawley Martin

You were born for a reason, and if you don't know the reason, you cannot be making progress toward its fulfillment—and so you likely feel that something's missing in your life. This book provides a methodology to learn why you're here, and it will hand you three keys that will open your eyes to what can bring you lasting happiness. Don't miss it.

Kindle: ASIN: B0C67MHNJF
PB: ISBN-13 : 978-1892538710

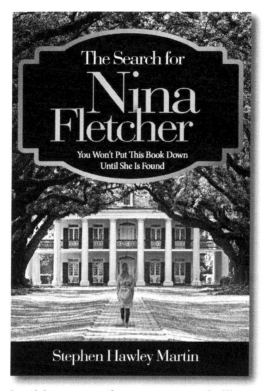

In this romantic suspense thriller, Rebecca wants to save the beautiful plantation home where she grew up, but to do so she must find her mother. If only she could remember what happened in the basement of the old house in Baltimore long ago. She must find out what happened there, she must!

Kindle: ASIN: B01J6MQZXS
Paperback: ISBN-10: 1535580879

Death in Advertising

FICTION FIRST PRIZE
WINNER
WRITER'S DIGEST

Stephen Hawley Martin

This whodunit set in an ad agency won First Prize for Fiction from *Writer's Digest* magazine. According to Mike Chapman, Editor-in-Chief of *ADWEEK* magazine, this novel is "A thrilling and evocative read. Masterful attention to detail brings the ad agency world to life and delivers a gripping whodunit." Get ready. You won't be able to put it down.

Kindle: ASIN: B00UIGGKUA
Paperback: ISBN-10: 1511662921

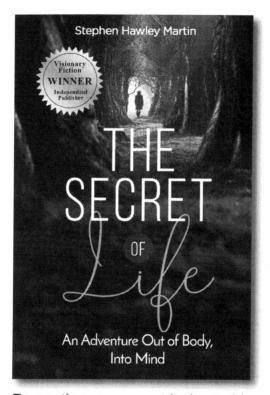

Romantic suspense at its best, this fast-paced novel won First Prize for Fiction from *Writer's Digest* and First Place for Visionary Fiction from *Independent Publisher* for good reason: It's very hard to put down. You'll be riveted as Claire flies to the island of Martinique to solve a mystery and soon realizes she's being stalked by a drug lord.

Kindle: ASIN: B08S7MG4WM
PB: ISBN: 979-8591416515

Printed in Great Britain
by Amazon

32353665R00084